VGM Opportunities Series

OPPORTUNITIES IN PLUMBING AND PIPE FITTING CAREERS

Patrick J. Galvin

Foreword by
Joe A. Childress
Executive Vice President
National Association of Plumbing-Heating-Cooling Contractors

VGM Career Horizons
a division of *NTC Publishing Group*
Lincolnwood, Illinois USA

Cover Photo Credits:
Front cover: upper left and upper right,
Plumbing and Mechanical magazine photos;
lower left, *Supply House Times* magazine
photo; lower right, Patrick Galvin photo.

Back cover: upper left and upper right,
Supply House Times magazine photos; lower
left, *Plumbing and Mechanical* magazine
photo; *SNIPS* magazine photo.

Library of Congress Cataloging-in-Publication Data

Galvin, Patrick J.
  Opportunities in plumbing and pipe fitting careers.

  (VGM opportunities series)
  1. Plumbing—Vocational guidance. 2. Pipe-fitting—
Vocational guidance. I. Title. II. Series.
TH6130.G34   1988   696. '1 '023   88-60906
ISBN 0-8442-6187-4
ISBN 0-8442-6188-2 (pbk.)

Published by VGM Career Horizons, a division of NTC Publishing Group.
© 1989 by NTC Publishing Group, 4255 West Touhy Avenue,
Lincolnwood (Chicago), Illinois 60646-1975 U.S.A.
Library of Congress Catalog Card Number: 88-60906
Manufactured in the United States of America.

8 9 0 BC 9 8 7 6 5 4 3 2 1

ABOUT THE AUTHOR

Patrick J. Galvin has been one of the most prolific writers in the plumbing, heating, and cooling industry for nearly 40 years.

In this time he invested 25 years as field editor, editor, and publisher of *Kitchen & Bath Business* magazine. He also has written for all national plumbing and heating magazines—and still does. In addition, he has authored one book for builders and architects and several books for consumers on kitchens and bathrooms.

He has conducted seminars for kitchen and bath dealers and distributors, has been a prominent speaker in the field internationally, and has been a leading consultant for foreign manufacturers looking to enter the U.S. market.

He is a graduate of the University of Illinois. He served in the Chinese Nationalist Army in World War II and was decorated by both the Chinese and U.S. armies.

ACKNOWLEDGMENTS

Thanks to the hundreds of plumbers I have known and interviewed over the last 40 years.

Special thanks to Joseph Borgia, apprentice coordinator for the Mercer County Vocational-Technical School, Trenton, New Jersey, for his generous sharing of time and knowledge.

Special thanks also to Bill Chamberlin, president of Chamberlin Plumbing & Heating Company and president of the New Jersey Plumbing-Heating-Cooling Contractors Association.

Thanks as well to the organizations that helped make this book valid: the National Association of Plumbing-Heating-Cooling Contractors; the Mechanical Contractors Association of America; the American Fire Sprinklers Association; the National Fire Sprinkler Association; and the National Association of Trade and Technical Schools. And finally thanks to Joe Mulrine, editor of the *U.S. Fire Sprinkler Reporter.*

FOREWORD

You're a young person thinking about your future. As a high school or college student, you've talked about employment and/or professional opportunities with guidance counselors and you've probably read material that outlines, in glowing terms, the way to make your fortune in any one of dozens of fields.

However, the counselors didn't have the time or the knowledge to go into detail about these opportunities, and the outlines were just that, outlines. They didn't provide the detailed information you need to make an intelligent decision.

You hold in your hands a book that provides the information you need to make an intelligent decision about a plumbing and pipe fitting career. The author, Patrick Galvin, has a background in the field, and he took the time to do an outstanding research job. These pages provide the information you need in complete, but easy to understand, terms.

You may be asking yourself, "Do I want to be a plumber and/or pipe fitter?" Perhaps you've thought about becoming a lawyer, doctor, or accountant or making a career in some other profession you perceive as offering a prestigious high-paying future. As you read *Opportunities in Plumbing and Pipe Fitting Careers*, you will learn that this field too offers a bright future. Trained plumbing and pipe fitting mechanics can count on steady work and exceptionally good pay. Perhaps, more importantly, that's not the end. Almost all of the contractors who are members of the National Association of Plumbing-Heating-Cooling Contractors are former mechanics. These men and women are truly professionals who have the respect of their friends and neighbors. Many are college graduates who in some cases went through apprentice training after getting degrees in such disciplines as engineering or business administration.

The plumbing-heating-cooling industry is very large, employing more than 1,500,000 persons in the United States. Plumbing, heating, and cooling make it possible for people to live in any climate and within highly confined areas without danger to health and comfort. An ample supply of pure water is necessary to any community. Proper installation of systems to handle the water supply and protect it from contamination is the responsibility of the plumbing industry.

Plumbing and pipe fitting mechanics must have a combination of mechanical aptitude and the ability to solve engineering problems by careful thought and creativeness. There is enough variety in the industry to satisfy the interests of almost anyone.

Yes! *Opportunity* is the key word in the title of the book you are about to read. Remember, the author is not only an expert writer but also an expert in the field of plumbing and pipe fitting. If you are ambitious and apply yourself, you will be welcomed to the industry. Once properly trained, you will be in demand.

Joe A. Childress
Executive Vice President
National Association of Plumbing-
    Heating-Cooling Contractors

INTRODUCTION

OK. It happens to everybody. Sooner or later you have to make a career choice. You have to decide what you will do for the rest of your life. Actually, it isn't that final. Years later you can change your mind, as many people do. But that's beside the point. The real point is you have to do something *now*.

Maybe you are doing it for the first time. Or maybe you already are in some other career, and you have to make a change. Regardless, the point is that you have to do something, and you have to do it right now.

What?

Plumbing? Be a plumber? Or a pipe fitter or steam fitter? Or a sprinkler fitter? What kind of a life is that?

A pretty doggone good one, according to 500,000 able and capable men and women who right now are paying for their condos and their Grand Ams or BMWs by fixing other people's pipes in other people's buildings—doing the jobs that a lot of other people, however smart they may be, just aren't good enough to do.

Well, that's good enough for a start. But think: What can it really mean?

First, imagine a colony on the moon or in space. Who, among all the high-tech specialists that would be involved in such a project, is going to be standing there with a pipe wrench ready to fix the interplanetary faucet? Who is going to hook up the toilet?

The answer is easy and obvious: A plumber.

Or think of a distant day when the oil from the Middle East dries up or is cut off. Who is going to do the piping for the energy that comes from our domestic oil, or from geothermal energy deep in the bowels of the earth, or from the sun or the sea? Yup. The plumber.

So if you are at a starting point or a turning point in your life, and if you want to do something that is both satisfying and productive, something that offers the opportunity either to work or to go out on your own, careers in plumbing or pipe fitting are worth looking into.

CONTENTS

xi

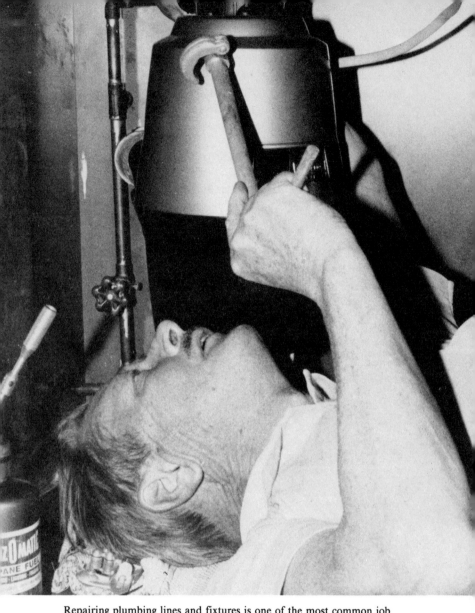

Repairing plumbing lines and fixtures is one of the most common job duties for plumbers. (photo courtesy of Patrick Galvin)

CHAPTER 1

THE PIPE TRADES

Any mention of "the pipe trades" usually will conjure up a vision of a plumber with a wrench, working under a kitchen sink. And in the real world, that very often will be exactly what a plumber will be doing and precisely where he or she will be doing it.

But that domestic plumber represents only a tiny fraction of a vast and varied industry.

The pipe trades are worked by such specialists as plumbers, pipe and steam fitters, sprinkler fitters, and many other subspecialists who might work on the sophisticated complexes that go into an oil or chemical refinery or a NASA launch site for space shuttles—or even the living and working complexes in space that will surely be part of our future.

On earth or in space, in almost any living or industrial situation, there always has to be a way to get supply or waste fluids and solids and gases from one place to another. Some are hot. Some are cold. Some require high or low pressure. So the pipe trades also work on simple and complicated systems and devices for relief of both pressure and temperature.

PLUMBING IN THE HOME

Consider, for example, the simplest and most widely recognized application of piping, the plumbing system in the home. The simple act of turning on a faucet in the kitchen or bathroom initiates a long and complex series of piping functions. The water that flows from the faucet must be drawn from an outside source, a city water system or a well, and a city water treatment plant itself must get the water from a river or reservoir that might be many miles away. The water must be delivered to the home under enough pressure to get it there over hill and dale, regardless of distance. It must also be delivered in quantities sufficient to supply the faucet even though a washing machine and a dishwasher might be demanding water from the same incoming supply at the same time, as might some family member taking a shower elsewhere in the house. Furthermore, the same or a similar collection of events will be taking place simultaneously in thousands of other houses served by this same water supply.

The incoming water must be distributed throughout the house to serve the kitchen, the bathrooms, and the laundry room through a system of metal or plastic pipes, but the system must be split to deliver both hot and cold water simultaneously, so that the person using the faucet can mix the hot and cold water to a desired temperature.

Then—Getting Rid of It

But that's only half of it. That water, once used, must also be carried away through a drainage system in the house and then through another outside the house to either a municipal sewer plant or a septic tank.

The in-house system works by gravity flow, so the pipes are fairly large. But they can't be too large, or moving waste would cling to the sides, where it would build up and cause stoppages that would have to be cleaned out. In other words, correct sizing of the pipe enables it to clean itself.

The drain pipes also must be vented to equalize air pressure inside the pipes. Without venting to equalize the air pressure, neither liquid nor solid waste could flow, and back-siphonage could cause the waste materials to back up in sinks and the toilet. To put it another way, the flow of waste materials confined in the pipe would cause a vacuum in its wake, so venting is needed to provide the make-up air to prevent such a vacuum by equalizing the air pressure.

So, obviously, high technology and science are involved even in this relatively simple domestic aspect of the plumbing trade.

And that is only part of the plumber's work in the home. There also will be gas lines that must be brought in safely, distributed to gas appliances, controlled against failure, and vented to the outside. There must be ductwork or piping for heating and cooling, and some of the water must be heated and distributed to the kitchen, to bathrooms, and to water-using appliances.

Consider how much more complicated the piping and plumbing can be in such commercial applications as a high-rise apartment building, a chemical processing plant, or a single oil refinery with its hundreds of miles of pipes.

SUN, SEA, AND SPACE

Look even farther ahead to the changing face of energy all over the world. Whatever its status right now, there surely will be solar energy in our future, and it will depend on piping and plumbing for effective use. There will be both energy and desalinated water from the oceans, and their effectiveness will depend on piping and plumbing. We already are using geothermal energy, and it too depends on piping and plumbing for effectiveness and efficiency. Piping and plumbing are visible in all aspects of a space launch, and these applications already are here. They will be equally important as human beings move communities and industries into space.

In the sprinkler fitting specialty of the pipe trades, every fire that results in loss of life leads to more laws requiring adequate protection. All over the country, sprinkler systems are mandated by law in homes, offices, and factories.

A VARIETY OF SPECIALISTS

In past years all of the workers who did all of these kinds of work were plumbers. They still are, but the tremendous growth of the nation and its change to an industrialized society have made it necessary for workers to specialize.

Pipe Fitters

Pipe fitters specialize in the installation, maintenance, and repair of industrial, commercial, and municipal high- and low-pressure pipe systems. They install the pipe lines to carry air, gas, oil, steam, or other materials.

Pipe fitters who work with steam are called steam fitters. Steam fitting is different from normal pipe fitting because of the high pressure of the steam.

Sprinkler Fitters

Sprinkler fitters perform a newer specialty, laying out and installing the fittings of a sprinkler system in an effective pattern to extinguish fires automatically in homes, hotels, or other commercial buildings. There are other sprinkler fitters who plan and lay out sprinkler systems for lawns and gardens. This work might not be as complicated or responsible as control of fires, but nevertheless it is one of the things sprinkler fitters do.

Plumbers

Plumbers generally do three kinds of work, and they might specialize in any of these three fields.

Possibly the biggest and steadiest of these three fields is the installation of plumbing lines and fixtures in new homes and the repair of such lines and fixtures in home remodeling, in which case plumbers usually work for plumbing contractors. Some specialize in new homes, others in remodeling.

Other plumbers lay water, gas, and sewer lines from supply mains to homes or other buildings, and they often work for utilities, developers, governmental units such as school districts, or builders. Still others might specialize in the water, gas, sewer, and vent lines in commercial buildings. They might work for larger mechanical contractors or for the building owners.

Heating, Venting, and Cooling Specialists

There are other subspecialties in the field of plumbing and pipe fitting. They involve heating, venting, and cooling buildings. Contractors who work in these subspecialties are generally known in the trade as HVAC contractors, for Heating, Venting, and Air-Conditioning.

Heating specialists might work in all types of heating, or they might work only with warm air heating or hot water (hydronic) heating.

A warm air specialist must be familiar with gas or electric furnaces and heat pumps, which supply the heat, and with the sheet metal ductwork that distributes it around the building.

A hydronic specialist must be familiar with boilers, which heat water, and the piping system that distributes the heated water around the house. The piping system might take the heat to radiators, or the pipes might be coiled in a ceiling or floor, or even be embedded in a concrete slab to heat by radiation. In some cases the boiler has a separate coil through which another water supply is routed to supply the building with hot water for sinks, tubs, and showers. This is an entirely different water system and requires separate piping.

Venting involves equalization of air pressure in plumbing systems, but it has many other aspects.

In homes it might involve evaluation and installation of such recently developed equipment as air-to-air heat exchangers to relieve a radon problem or a problem of formaldehyde emissions.

These problems were unknown a decade ago. Radon is a natural gas that comes from decay of natural uranium. It enters a home from the soil under it. Formaldehyde is used in many products in the home, including furniture and paneling, but some people are allergic to it. The problem of allergy to such indoor conditions didn't exist a few years ago when houses were built to "breathe" and indoor air pollution was exhausted routinely. The problem showed up and was aggravated in the supertight, highly insulated homes of the 1980s.

In commercial buildings and factories, venting can involve the exhausting of various types of pollution and fumes, and venting must supply fresh air to replace the air that has been exhausted. This is known as "make-up" air.

Air-conditioning, like heating, changes the climate in a building to make it more comfortable. A cooling system often is "piggybacked" on the heating system, using the same piping or ductwork. At other times a cooling system requires new piping or ductwork. In any case, the air-conditioning specialist needs to have a knowledge of refrigeration theory and principles and of the way the compressors and condensers that refrigerate the air or water for distribution interact.

JOBS, METHODS, AND MATERIALS

Obviously, piping systems will differ by size and operation, according to their purpose.

In homes, for example, a water system will be of copper, plastic, or galvanized steel, and these can be handled and installed by one or two workers.

But a municipal sewer system uses large clay pipe that requires a crew of several workers. A municipal water supply pipe can be so large that a person can walk upright in it.

Skilled plumbers and fitters must know the characteristics of all of these materials, when and when not to use any one of them, and how to cut, bend, and fit them together. All plumbers and fitters must be able to read blueprints and follow building plans and instructions of supervisors.

They also must be aware of national and local building codes that are written to protect the health and general welfare of the public, since their work relates more directly to public health and sanitation than that of any other construction trade. The codes sometimes restrict use of specific materials in one area that might be acceptable in another neighboring community. Plumbers and fitters always must be alert to such changing requirements and conditions.

Plumbing and pipe fitting work is demanding physically and challenging mentally. (*Plumbing & Mechanical* magazine photo)

CHAPTER 2

A CAREER FOR YOU?

Do you want to be a plumber? If your answer is "Yes" or "Maybe" there is good reason to investigate the issue further. Consider these advantages:

1. It is one of the highest paying of the skilled construction trades.
2. The fringe benefits are as good as, or better than, any others.
3. It is popular, and it is growing in scope. But it is not so overcrowded that you might find it hard to get a job.
4. It is progressive, with advancing technology that keeps it challenging and interesting, and with many opportunities for advancement.

HIGH PAY SCALES

Plumbers and pipe fitters held 408,000 jobs in 1986. Their pay and fringe benefits varied in line with negotiated union and nonunion contracts. Wages and fringes in some typical contracts in 1986 included $24.20 per hour in St. Louis, $22.95 in Chicago, and $20.94 in Peoria.

For those working in large organizations, a second shift normally will gain a differential of about 10 percent, and a third shift (night work) will gain a differential of about 15 percent. In areas where there is a lot of construction, a lot of overtime work might be available at premium pay rates of 50 percent or 100 percent over the regular hourly scale.

In 1986 median weekly earnings for plumbers and fitters was $470. Most earned between $330 and $614 weekly. The highest 10 percent earned $774 weekly; the lowest 10 percent earned about $246 weekly.

In 1986 hourly wage rates for maintenance pipe fitters in metropolitan areas was $14.35. In comparison, average wage for all nonsupervisory and production workers in other trades was only $8.75. Wage rates in the Midwest and West tend to be higher than in the Northeast and the South.

First, of course, it is necessary to learn the trade. This is done through an apprenticeship, which we will cover in detail in chapter 4. But the pay for starting apprentices usually is 40 to 50 percent of the wage rate paid to experienced journeyman plumbers or fitters. The pay increases periodically as skill improves, and after the four-year apprenticeship the beginner moves up to full-scale pay and benefits.

GOOD JOB OPPORTUNITIES

The federal Bureau of Labor Statistics predicts about 19,000 job openings per year through 1995. This can increase dramatically if there are technological breakthroughs in such areas as coal gasification, nuclear power plants, or alternate forms of energy, or if there are changes that lead to construction of more oil refineries. The demand for piping, some of it very complex and involved, is tremendous in all of those industries.

Construction plumbers who work in new homes sometimes are subject to seasonal layoffs. But in recent years residential remodeling has grown bigger than new construction, and remodeling continues strong through the winter months. So, winter or summer, the probability of continuing employment is almost assured.

FRINGE BENEFITS

Fringe benefits for plumbers and fitters are generally in line with those of other skilled construction trades. These include paid vacation, sick leave, a specified number of holidays per year (generally about 13), medical and hospitalization insurance, bonuses, and pensions. All of these, of course, depend on employment agreements between the employer and the workers. In union situations they are part of the contract. In nonunion situations they might be standard practice for the employer, or they might be negotiable on an individual basis.

CHALLENGES FOR YOU

This is not easy work. It is demanding physically and challenging mentally. A plumber might work one day in a basement and the next day in a high-rise apartment building. The work might be inside in relative comfort or outside in snow, rain, or frigid conditions. Working space often will be cramped, and it will take a lot of muscle to maneuver plumbing fixtures or large pipe into place. There always is some danger when working with torches for soldering pipe or with cutting tools, and burns are possible from steam or hot pipes. The danger in working with gas lines is obvious. In this field, however, most accidents are more of a nuisance than they are serious.

There always are intellectual challenges. In either new home construction or remodeling, the plumber comes twice and spends two different periods on the job.

The first job is rough-in. After a house or other building is framed, with the floor joists and wall studs in place, it is the plumber's job to install the pipes and fittings that go behind the walls or under the floors.

The second job at the same site will be the finish work, after the carpenters, plasterers, and tile setters have finished off walls and floors. At that time all fixtures must be hooked up with the pipe connections that were prepared in the rough-in stage.

All of this can be relatively easy in new homes where there is ample working space and no home owner looking over the worker's shoulder. But in remodeling, there is no way to know what might be encountered when an old wall is opened up. Plumbers expecting an easy rough-in might find remnants of a forgotten chimney in the wall where a water supply pipe is supposed to go. There might be a hidden duct for air conditioning where the remodeling plan calls for a waste or vent line.

So the plumber needs ingenuity in coping with unexpected problems. Sure, it always is easy to call the employer for an answer, but the plumber who can solve field problems quickly and easily with brainpower will be the one who gets premium pay and promotions.

Even more exciting are the frontiers of geothermal energy, which comes from tapping the heat deep in the earth; desalination of seawater to provide supplies of fresh water from this inexhaustible source; progress in photovoltaics to make energy from the sun more generally usable; and the prospects for manned stations and communities in space.

All of these translate into vast needs for extensive, complicated piping complexes. The job market and opportunities for plumbers and pipe, steam, and sprinkler fitters can only increase in years to come. And, it should be noted, the biggest personnel shortages right now are not in the journeymen

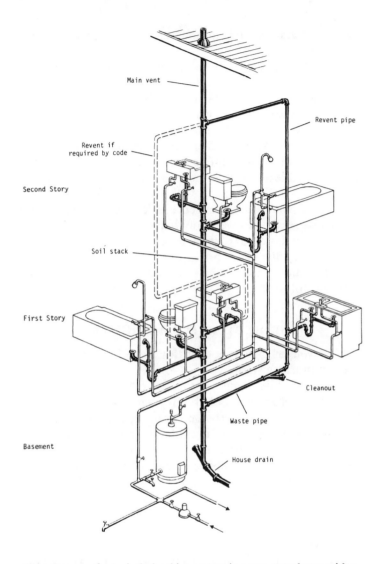

Main vent

Revent pipe

Revent if
required by code

Second Story

Soil stack

First Story

Cleanout

Waste pipe

Basement

House drain

This diagram of a typical plumbing system in a two-story house with a basement shows the waste stack vented to the roof and all connections to the kitchen, laundry, and two bathrooms.

who cut and lay the pipes but in management—in the field and in the office.

For any of you who think big, plumbing is the largest single industry in the United States and one of the most important. There is no segment of an industry more challenging than that of management, whether it involves managing a project in the field or from an office. This demands a professional person capable of coordinating and supervising all aspects of a job and who is equipped with comprehensive technical knowledge, sound business sense, strong leadership ability, and a progressive, forward-looking attitude.

Each year hundreds of managerial positions go unfilled in this industry due to a shortage of qualified applicants.

So the opportunity is there. It waits only for someone to take control.

MORE THAN EQUAL RIGHTS

Fine, you say, but what does this mean for women or any other minority group?

In the plumbing trade, your rights are more than equal. In 1978 new Department of Labor regulations went into effect that require apprenticeship program operators to take affirmative action in recruiting and hiring women.

This means goals and timetables must be established to make sure that such groups as women, blacks, and Hispanics are represented in the plumbing trade in accordance with their numbers in the labor force in the community. These rules or similar state laws apply to all apprenticeship programs registered with the Labor Department's Bureau of Apprenticeship and Training or a recognized state apprenticeship agency. In addition, the Civil Rights Act and state laws require programs, including unregistered ones, to provide equal opportunity in employment.

The regulations clearly outline how an apprentice program sponsor is to decide whether the program underutilizes these groups. Programs found to underutilize these groups must actively go out and try to recruit them. Every program that has at least five apprentices must have a written plan that sets goals for increasing the number of women and minorities, with a timetable to reach these goals.

Many people learn the trade by hiring on as an assistant with a local contractor. (*SNIPS* magazine photo)

CHAPTER 3

EDUCATION AND TRAINING

Most plumbers and fitters learn their trade through an apprenticeship program (see chapter 4). This procedure takes four years officially. In union programs, however, it recently was extended to five years, because unions judged that a higher level of craftsmanship was needed due to the greater size and complexity of the industry.

Employers and unions prefer persons who are at least 18 years old and who have a high school diploma or equivalent. They also must be in good physical condition, since the work can be hard.

Although the trade is highly unionized, it is not necessary to join a union or to take a union apprenticeship program. Anyone can go through an apprentice program and win a certificate either within the union framework or totally outside of it. Several different agencies cooperate with the needs of both the industry and individuals to sponsor apprenticeship programs. Some cost money. Some are free.

The costs of any professional courses such as these might be borne by, or shared with, your employer. If you are just starting out and an employer likes your work, he or she probably will want you to go through an apprenticeship program. So be prepared to bring it up and to discuss it.

Many persons learn the trade on the job by hiring on as a

plumber's helper with a local plumbing contractor. While this provides an opportunity to work and to learn, the experience and learning will be limited to the kinds of work done by that particular contractor. For example, this plumbing contractor might do only home repair work and never get into heating, sheet metal, or any commercial work, and all of these involve skills that any plumbing journeyman is expected to know.

On-the-job training should be supplemented by classroom work at a regional vocational school or community college that offers such training in an approved apprenticeship program. There are many of these in nearly all states. This training also is available in many communities all over the country through union facilities. The primary union in this field is the United Association of Journeymen and Apprentices of the Plumbing and Pipe Fitting Industry of the United States and Canada, usually referred to in the trade as the UA.

Keep in mind that apprentice programs require an average of 144 hours of classroom work per year, plus about 2,000 hours per year of actual job performance in the field.

TRADE AND TECHNICAL SCHOOLS

It also is possible to get training at an accredited trade or technical school. These are full-time schools, so it is possible to become educated in the plumbing trade in two years (plus the required field experience), depending on how fast and hard you want to work. Technical schools that offer plumbing, however, are limited. Of the hundreds of schools listed in the 1987 handbook of the National Association of Trade and Technical Schools, only 12 offer plumbing. And since these do not include the on-the-job training required in an apprenticeship program, they do not lead to certification as an apprentice.

As our society becomes more and more conscious of credentials, lack of an apprenticeship certificate might prevent you from finding the job you want.

HOME STUDY COURSE

For those who might be too distant from a vocational or technical school, a home study course is available from the National Association of Plumbing-Heating-Cooling Contractors (NAPHCC). This course is approved by the national Bureau of Apprenticeship and Training and is set up to take four years, but it covers the same material used by instructors teaching the five-year apprentice course at vocational centers operated by the UA.

The NAPHCC home study course includes a book of lesson plans for each year and a student workbook. These are written simply so they can be understood easily, with no technical jargon, and they are fully illustrated. But if a student should have any problem with the course, the answer is as near as the telephone. A coordinator is always available at NAPHCC to help students with problems.

NAPHCC suggests that a student take 45 days to study a lesson and return the completed test. At NAPHCC the test will be graded, and comments will be made to help the student in future lessons. After all lessons and tests have been completed successfully, there is a final examination for the year, and a certificate is awarded recognizing the level of accomplishment.

At the end of the fourth year there is a final graduation certificate.

Any student who is taking the home study course should, preferably, also be working for a licensed plumbing contractor at the same time. Where this is impossible, a student still may take the entire course, but job training will be necessary at some point for certification as an apprentice.

Cost of the home study course is $500 per year for students working for a member of NAPHCC and $700 for others. It includes copies of the *American Society of Sanitation Engineers Dictionary,* the *NAPHCC Safety Manual,* and a first aid manual. At the start of the second year of study a copy of the *National Standard Plumbing Code* is sent to the student. These reference books are to help the student in any field work he or she might be experiencing while taking the course.

For more information on the home study course, write to NAPHCC, 180 South Washington Street, Falls Church, VA 22046.

SCHOLARSHIPS

NAPHCC also offers a scholarship plan for plumbing students. Three scholarships are available each year. To compete for one, you must be sponsored by a NAPHCC member and must be in your final year of high school or freshman year of college. You must apply for a scholarship before May 1 of any given year. Each scholarship pays $2,500 annually for up to four years. The money goes directly to the college.

In addition to this national scholarship, the NAPHCC Women's Auxiliary and many state and local PHCC associations have scholarship programs.

For more information on a scholarship write to NAPHCC Educational Scholarship, PO Box 6808, Falls Church, VA 22046, or call (800) 533-7694.

NAPHCC APPRENTICESHIP TRAINING PROGRAMS

NAPHCC supports the joint apprentice training programs for union apprentices and also provides a four-year NAPHCC plumbing and HVAC apprenticeship training program administered through state and local NAPHCC contractor committees. The committees operate privately or in conjunction with community or technical schools. This program is sponsored statewide in several states. NAPHCC's curriculum philosophy is to develop a well-rounded mechanic. This is accomplished through stressing the "how-to-do-tasks" philosophy and by explaining the background and theory of the task. The NAPHCC curriculum builds on and reemphasizes key points from year to year. This type of training is the opposite of modularized task training programs, in which the apprentice sees the task within one module and receives limited theory explanation.

SPRINKLER FITTING SCHOOLS

For sprinkler fitters, further education is provided by the American Fire Sprinkler Association (AFSA).

Since sprinkler fitting is a specialty within the plumbing and pipe fitting industry, AFSA courses are designed to further your education in this specialty rather than general plumbing or fitting.

Several courses and seminars are offered. A typical course on "Principles of Fire Sprinkler Design" takes three weeks and costs $1,250, or only $1,000 if you are employed by a member of AFSA. This course, given regionally across the country on different dates, covers such topics as economics of design function, construction types, dimensioning, elevations and slopes, materials, maximization and minimization, and management by objective. The course enables the student to become familiar with sprinkler system components, such as piping, fittings, valves, and hangers; how these components fit together; and the basic application rules of fire protection sprinkler systems. It also features design exercises.

Typical seminars are held for one or two days and cover such subjects as fire pumps, hydraulic calculations, residential systems, and standpipes. Seminars also are presented regionally, and they cost about $150 per day, less for AFSA members.

For information write to AFSA Seminars, 11325 Pegasus, Suite E-109, Dallas TX 75238, or call (214) 349-5965.

WHEELS OF LEARNING

Apprenticeship is only one of several phases in the *Wheels of Learning* (WOL) educational program of the Associated Builders and Contractors, Inc. This association of about 20,000 construction contractors, including plumbing contractors, has as its mission the promotion of what it calls the "merit shop." ABC is not necessarily anti-union, but it definitely favors open shops and the awarding of construction contracts on the merit of the contractor rather than on whether the contractor is unionized.

A Wheels of Learning educational program can be a conventional apprenticeship as already described, and this quite possibly is best for a new trainee in the field. But it also can be "task-oriented"—that is, it can include whatever courses, or modules, that are needed to perform specific kinds of work in any given project.

For example, one module for plumbers is "Joining Clay and Concrete Pipe." This is a usable, marketable skill that can be taught to a trainee in 2 1/2 hours, after which the trainee can immediately be productive on the job. The only restriction on this course is that the trainee already have taken one prerequisite module on "Basic Plumbing Tools," because the knowledge and skills in that module are necessary to successful completion of this one.

Task-oriented programs also can be very valuable for cross training. That means you, working and studying to be a plumber, can take special modules on pipe fitting that are more advanced than your plumbing apprenticeship would offer. Or you might want to get a job in maintenance at a big hotel-office complex and find you need more electrical or HVAC skills. You could get them through these WOL modules.

WOL instruction can be in a classroom setting, in a contractor's shop, on a job site, or even on a one-to-one basis. ABC four-year programs related to the plumbing field include:

• Pipe Fitting I, II, III, IV
• Plumbing I, II, III, IV
• HVAC I, II, III, IV
• Sheet Metal I, II, III, IV

Other programs offered in other trades include electrical work, painting, masonry, carpentry, and welding.

Again, this training by ABC is task-oriented. That means that the modules differ from representative courses in other types of schools.

In chapter 4 you can read a full description of a plumbing apprentice course in a typical vocational-agricultural school. That chapter will also describe enough of the ABC plumbing modules so you can see the comparison.

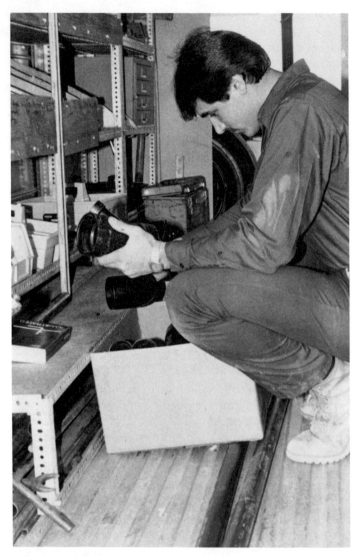

This apprentice is responsible for gathering the materials needed for the next day's jobs. (photo courtesy of Patrick Galvin)

CHAPTER 4

THE APPRENTICE PROGRAM

There are hundreds of good reasons why the concept of apprenticeship was devised for many crafts and trades. One of the best reasons is the one that governs the plumbing field, and that is that the safety and welfare of more than two hundred million Americans depend on good, safe water supply and waste disposal.

The same thing is true, of course, all over the world. We hear fairly frequently about disease sweeping a third world country and killing thousands of people. We don't often hear such disastrous news about countries where sophisticated systems have been set up to ensure safe water supply and organized, sanitary sewage disposal.

It is safe to say that there is ample justification for government involvement in apprentice programs. And the government is very much involved now, through the National Apprenticeship Act, passed by Congress in 1937. However, that was the era of the Great Depression, and actually the act was passed more to protect the rights of workers than for the public welfare.

ORIGINS OF APPRENTICESHIP

Apprenticeship dates back to the code of the Babylonian ruler Hammurabi, which required artisans to teach their skills to the young.

But modern apprenticeship as brought to this country from Europe has its roots in what used to be called indenture. Indenture involved a contract binding a worker to a skilled craftsperson in what was very near to a master-slave relationship, and in early New England the apprentice depended on the master for food, clothing, and living quarters.

As industry expanded after the Industrial Revolution, apprenticeship also was revolutionized to apply more to the new machine age. The early system of depending on the master for food, clothing, and lodging disappeared, and compensation was changed by employers to the payment of wages. The wages were small, but there was a predetermined scale so the apprentice made progress.

In some trades, however, the term *master* continues to this day. We have, for example, "master plumbers" and "master machinists."

APPRENTICESHIP TODAY

Today the National Apprenticeship Program includes the coalition of management, labor, and government that supports apprenticeship in the United States, along with all of the programs and the enrolled apprentices that implement the concept nationwide.

That means that all of those elements work together. Management means all of the various bosses, the owners of businesses, the big contractors, whoever might hire a plumber or fitter. Labor means the unions, not only the giant UA but also any smaller unions that might have grown up around the country. Government means Uncle Sam is watching over it all, setting out regulations to govern apprentice programs and approving or rejecting programs that might be advanced by schools or unions or bosses.

And it means that you can have something to say about it also. If you see that you need some special training in the plumbing or fitting field, you can go to those who run the training in your area and discuss it.

Apprenticeship programs are operated by employers, employer associations, schools, and labor unions. The government role is to provide support services for these program sponsors.

Under the National Apprenticeship Act, the Bureau of Apprenticeship and Training (BAT) is responsible for providing service and technical assistance to existing programs and to those who want to set up a program. To do this, the bureau works with and through state apprenticeship councils.

HOW A TYPICAL PROGRAM WORKS

For a look at how the National Apprenticeship Program can work for you, let's drop in for a visit at a typical vocational-technical school. The one we're going to look at is the Mercer County Vocational/Technical School (MCVTS). Located in green fields near Trenton, New Jersey, it is one of three such schools in the county. Only one county in the state has no such school, which is a good indication that you probably can find one near you, wherever you are.

Here we meet Joseph Borgia, the apprentice coordinator for the vo-tech schools in the county. He works with an advisory committee drawn from building inspectors, contractors, teachers, and students, and this group continually reviews what is taught and considers what should be improved or changed.

As an example, a year ago the course included 36 hours on lead wiping. This is the complicated technique of joining cast-iron pipe. An area plumbing inspector told Borgia that this was no longer done in the area, and some contractors pointed out that there was instead a need for education and training in hydronic heating. In hydronics the heat is generated in a boiler and then distributed by hot water.

So hydronic heating replaced lead wiping in the curriculum.

How It Works for You

Let's assume you're going to take the apprentice course at MCVTS. The course will last for four years, and in that time you go to school two nights a week, three hours each night. You must complete 144 hours of school each year for a total of 576 hours in the four years. That means 24 weeks of school each year spread over about eight months, and it roughly follows the regular school year.

But that is only part of it. In the four years you also must complete 8,000 hours of on-the-job training before you get a graduation certificate and become a journeyman. The on-the-job training will be accomplished as you work for those four years for a plumbing contractor in your area. That contractor will be your sponsor in the apprentice program.

If you have no sponsor you still can take the training course on your own. But you will not get your certificate until you have completed 8,000 hours of work on the job. That doesn't mean you can't get a job. It does mean that you probably won't get journeyman pay if you work for a union shop. If you work for a nonunion shop, on the other hand, and if you are good at your work, you might get paid a lot more than journeyman scale.

The union believes four years is not enough and in 1987 expanded the time to five years. So if you go to a union school instead of a vo-tech school it will take longer—unless the government changes the requirement.

In the union view, it now takes more than four years to bring an individual to the required level of craftsmanship, because the field and its technology are becoming much more complicated. In a five-year program the first year will be a probationary year in which you explore the various specialties of the trade in more depth before you actually lock in on it.

What, you might wonder, would you study for all of these years in school? Following is the complete class period breakdown for the plumbing program at MCVTS, and it is typical. Each block of instruction consists of 36 class hours and is broken down into 12 three-hour class sessions.

FIRST YEAR

A. Basic Industrial Math

1. Adding and subtracting whole numbers
2. Multiplying and dividing whole numbers
3. Adding and subtracting fractions
4. Multiplying and dividing fractions
5. Adding and subtracting decimals
6. Multiplying and dividing decimals
7. Percentage problems
8. Ratio and proportion
9. Squares and square roots of numbers
10. Area and perimeter
11. Right triangles—area and degrees
12. Area and circumference of circles

(This block is designed to enable you to do the basic mathematical operations necessary to function in the plumbing trade.)

B. Basic Blueprint Reading I

1. Introduction, orientation, use of equipment
2. Reading basic scales: full, half, and others
3. Introduction to orthographic projection
4. Principles of orthographic projection
5. Advanced orthographic projection
6. Basic sectioning techniques
7. Offset, partial, rotated sectioning techniques
8. Utilization of the engineering drawing
9. Introduction to dimensioning practices
10. Basic decimal dimensioning practices
11. Basic fractional dimensioning practices
12. Concepts of isometric drawings

(This block is designed to enable you to interpret basic drafting data from engineering drawings related to plumbing.)

C. Math II for Construction Trades

1. Review of measuring; use of carpenter's ruler
2. Review of fractions; board feet
3. Review of decimals
4. Review of square roots
5. Pythagorean theorem; 3-4-5 triangle
6. Pythagorean theorem application
7. Rectangular volumetric measurement
8. Volumetric measurement; concrete calculations
9. Cubic yards vs. cubic feet vs. cubic inches
10. Introduction to trigonometry and angular measure
11. Trigonometry applied
12. Layout and estimating applied

(This block is designed to enable you to do the mathematics to solve basic linear, volumetric, and area problems.)

D. *Blueprint Reading II, Construction Trades*

   1. Architectural and civil engineer's scales
   2. Site plan interpretation
   3. Floor plan design considerations
   4. Floor plan development
   5. Window plan dimensioning
   6. Window and door schedules and delineation
   7. Footing and foundation drawing development
   8. Footing and foundation layout
   9. Footing and foundation dimensioning techniques
  10. Wall section elevation layout
  11. Wall section dimensioning techniques
  12. Exterior wall elevations

(This block is designed to enable you to interpret construction and materials information from a set of architectural drawings.)

SECOND YEAR

E. *Plumbing—Related Theory I*

   1. Code construction, 22 basic principles
   2. Code construction II
   3. Code interpretation I
   4. Code interpretation II
   5. Code applications I
   6. Code applications II
   7. Code applications III
   8. Code applications IV
   9. Code applications V
  10. Code applications VI
  11. Code applications VII
  12. Code applications VIII

(After this study you can interpret sketches often used as

part of the plumbing examination and can apply plumbing code content to residential structures.)

F. Plumbing—Related Theory II

1. Fitting and piping specifications
2. Sewage disposal I
3. Sewage disposal II
4. Water distribution, hot and cold
5. Terminology I, drainage
6. Terminology II, venting
7. Introduction to house sewer connections
8. Introduction to house drain connections
9. Fixture units, size, I
10. Fixture units, size, II
11. Installation procedures, fixtures
12. Types of water closets

(This block is designed to enable the student to discuss the theory of various installation procedures and use the correct terminology associated with such procedures.)

G. Plumbing Math I

1. Right angle with 45° diagonal
2. Right angle with 45° diagonal
3. 45° offset with wye fitting
4. 45° offset with wye fitting
5. Wye and tee wye assemblies
6. Wye and tee wye assemblies
7. .707 constant in 45° piping assemblies
8. .707 constant in 45° piping assemblies
9. Grade, drop and run
10. Grade, drop and run
11. Loops with grade
12. Loops with grade

(In this block you will use a basic transit and level to perform the mathematical calculations necessary to determine land elevation and grades in construction.)

H. *Plumbing Math II*

1. Solutions with 60° diagonals
2. Solutions with 22 1/2° diagonals
3. Solutions with 11 1/4° diagonals
4. 45° offsets in parallel
5. Special case of 45° and 90° fittings
6. Volumes
7. Water pressure head and force
8. Water measure
9. Elevations and grade
10. Elevations in plan view
11. Basic transit
12. Basic level

(After this 36 hours you can do calculations when using common offset fittings. You also can calculate force, pressure, and volume in various lengths and sizes of pipe.)

THIRD YEAR

I. *Plumbing—Drawing I*

1. Architectural symbols and graphics
2. Layout details
3. Introduction to pipe drawing
4. Introduction to job specifications
5. Basement floor plans, detail drawings
6. Kitchen floor plans, detail drawings
7. Rough-in kitchen sink, detail drawings
8. Bathroom floor plans, detail drawings
9. Rough-in lavatory
10. Rough-in water closet combinations
11. Rough-in bathtub with shower

12. Rough-in drawings for various fixtures

(With this you will be able to develop freehand layouts and detail the drawings required for plumbing permits according to code.)

J. Plumbing—Drawing II

1. Composition of a set of drawings
2. Specifications
3. Plot plans
4. Architectural plans
5. Mechanical plans
6. Floor plans
7. Structural plans
8. Elevation plans
9. Details
10. Sectional details
11. Plumbing plans
12. Plumbing details

(This block is designed to enable you to extract specific plumbing content from a set of construction drawings.)

K. Plumbing—Related Theory III

1. Review of theory II
2. Major topic paper, I
3. Major topic paper, II
4. Major topic paper, III
5. Major topic paper, IV
6. Plumbing traps, I
7. Plumbing traps, II
8. Plumbing traps, III
9. Venting, I
10. Venting, II
11. Venting, III
12. Venting, IV

(Here you can research one of 10 topics and present a 30-minute report to the class. Topics include background, history, and application of venting, drainage, fixtures, and traps.)

L. *Plumbing—Related Theory IV*

1. Introduction to indirect heaters
2. Relief valves—lever, spring types
3. Reducing valves
4. Thermostats
5. Conduction, convection, radiation
6. Side arm heaters
7. Heating devices
8. Storage tanks
9. Hot water circulation
10. Summer-winter hookup
11. Automatic storage gas water heaters
12. British thermal units (BTUs)

(This block will enable you to describe the operating principles of various types of heating units using common trade nomenclature.)

FOURTH YEAR

M. *Hydronic Heating I*

1. Introduction to hydronic heating
2. Principles of gas combustion
3. Principles of oil combustion
4. Components of gas-fired boilers
5. Components of oil-fired boilers
6. Basic electric circuit concepts
7. Low-voltage wiring and controls
8. Gas boiler/low-voltage controls
9. Gas boiler spark ignition, vent dampers

10. Oil boiler/low-voltage controls
11. Boiler loss calculations
12. Boiler sizing and selection

(This block enables you to explain the principles of gas-fired and oil-fired hot water boilers, including installation, combustion, and control circuitry.)

N. Hydronic Heating II

1. Hot water coils, types, characteristics
2. Radiators
3. Hot water baseboard systems
4. Expansion tanks, installation, function
5. Circulating pumps, zone valves
6. High efficiency gas boilers
7. Combustion efficiency test procedures
8. Hot water heaters
9. Mechanical troubleshooting gas boilers
10. Mechanical troubleshooting oil boilers
11. Heat loss calculations
12. Baseboard heat selection

(After this block you can determine the processes required to select and install water and steam heating fixtures and components and explain the function and operation of each.)

O. Principles of Process Piping

1. Pipe schedules and materials
2. Fittings
3. Flanges, types and pressure ratings
4. Pumps, centrifugal
5. Pumps, positive displacement
6. Chemical compatibility
7. Gasket materials
8. Valves, uses and selection

9. Valves, ratings
10. Vessels and tanks
11. Vessel heads and fittings
12. Welded systems

(This block enables you to describe uses, configuration, advantages, disadvantages for all the mentioned components in process piping.)

P. Isometric Drawing, Exam Terminology

1. Principles of isometric drawings
2. Isometric axis
3. Interpreting isometric projection planes
4. Drawing interpretation I
5. Drawing interpretation II
6. Drawing interpretation III
7. General exam content, structure
8. Basic terminology
9. Review of code requirements
10. Review of code requirements
11. Review of fixture installation
12. General review

(With this block you will be able to interpret and sketch basic isometric piping systems, and you will also become familiar with licensing examinations.)

As we noted before, the plumbing field has grown immensely over the years, and as a result, certain parts of it have become specialties on their own.

Therefore, to become a licensed pipe fitter you will have to take different courses after the first year and a half. As the program is set up at our typical vo-tech school, courses are the same for sections A through F. At section G, which comes midway through the second year, the subjects change.

You would enroll in either the plumbing or pipe fitting apprentice course. Following is the course breakdown for pipe fitting, except for the first 1 1/2 years. Since sections A through F are identical to the plumbing course outlined, refer to those sections. We don't repeat them here. Your course will continue with the following, again with each block representing 36 class hours broken down into 12 three-hour class sessions.

SECOND YEAR

G. *Testing, Analysis of Combustion Efficiency*
1. Chemistry and physics of combustion
2. Conditions necessary for combustion
3. Types of testing equipment
4. Draft gauge readings
5. Heat measurement and transfer
6. Combustion testing
7. Recording stack temperature readings
8. Smoke test readings
9. Inspection of flame and color
10. Products of combustion
11. Carbon dioxide testing
12. Maximizing BTU efficiency

(This will enable you to conduct a thorough combustion test on any domestic heating unit and determine its efficiency.)

H. *Vessel Installation and Piping*
1. Surface and submerged tanks
2. Gravity-fed, single pipe systems
3. Twin tank piping
4. Selecting pipe and fittings
5. Plumbing requirements and codes
6. Testing procedures
7. Installing surface and submerged tanks

8. Plumbing single pipe systems
9. Plumbing double pipe systems
10. Installing pipe thread fittings
11. Installing tubing and sweat fittings
12. Pressure testing installed piping

(This will enable you to vent and install the required pipe in various types of tank and boiler installations.)

THIRD YEAR

I. Plumbing Math III

1. Solutions with 60° diagonals
2. Solutions with 22 1/2° diagonals
3. Solutions with 11 1/4° diagonals
4. 45° offsets in parallel
5. Special case of 45° and 90° fittings
6. Volumes
7. Water pressure head and force
8. Water measure
9. Elevations and grade
10. Elevations in plan view
11. Basic transit
12. Basic level

(After this block you can perform calculations when using common offset fittings. You also can calculate force, pressure, and volume in various lengths and sizes of pipes.)

J. Welding Theory I

1. Introduction to welding
2. Personal safety practices
3. Machine and equipment safety
4. Oxyacetylene cutting equipment
5. Oxyacetylene cutting theory
6. Oxyacetylene welding equipment

   7. Oxyacetylene welding theory
   8. Shielded metal-arc welding equipment
   9. Shielded metal-arc welding theory
  10. Electrode functions and theory
  11. Recognizing weld defects
  12. Basic joint configuration and theory

(This block will enable you to identify and explain operation and safety procedures to operate oxyacetylene welding and cutting equipment, shielded metal-arc welding equipment, and the purpose of various electrodes, types of joints, and structural shapes used in the welding industry.)

K. *Cutting and Flat Welding*

   1. Basic cutting torch setup and adjustment
   2. Hand-held cutting, straight and bevel
   3. Hand-held cutting, piercing and structural shapes
   4. Set up and operation of automatic cutting machine
   5. Shielded metal-arc welding and machine setup and operation
   6. Linear beads using various electrodes
   7. Weaving techniques using various electrodes
   8. Flat fillet, single pass welds
   9. Flat fillet, multipass welds
  10. Flat fillet welds using weaving techniques
  11. Flat square butt welds
  12. Flat vee groove techniques

(This teaches you to demonstrate basic cutting and flat welding procedures and techniques.)

L. *Oxyacetylene Welding Techniques*

   1. Equipment setup and safety
   2. Selection of tips and supplies
   3. Running beads, rod at various positions
   4. Various position fillet welds

5. Welding heavy materials
6. Running brazing beads, various positions
7. Brazing fillets in various positions
8. Brazing heavy metals
9. Brazing cast iron
10. Brazing alloy materials
11. Soldering thin materials
12. Soldering dissimilar metals

(This teaches you to weld, braze, and solder various metals using oxyacetylene processes.)

FOURTH YEAR

M. Bench Work and Drill Press

1. Tool safety
2. Hand tool identification
3. Safety in soldering
4. Hard soldering and soft soldering
5. Use and care of hand grinders
6. Drill press safety
7. Drill press setups
8. Various drill chucks
9. Microscopic hole drilling
10. Reaming and drill press operations
11. Drilling feeds and speeds
12. Drill sharpening, angles, clearances

(After this block you will be able safely to perform hand tool bench operations, soldering, and basic drilling.)

N. Plumbing—Related Theory V

1. Review of all theory sections
2. Traps
3. Loss of seals
4. Inspection, testing plumbing systems

  5. Grease traps
  6. Sewage ejectors
  7. Plumbing symbols
  8. Plumbing fixtures
  9. Code review
  10. Preparation for lead work
  11. Complete coverage of text
  12. Final review of theory sections

(You will be able to correlate and apply plumbing content, mathematics, and basic skills for the plumbing examination.)

O. Fasteners and Threads

  1. Power threads
  2. Unified thread series
  3. Thread data tables
  4. Basic thread terminology
  5. Common hardware
  6. Hardware mating configurations
  7. Loads on bolts/nuts
  8. Loads on bolts/nuts
  9. Loads on bolts/nuts
  10. Specialized fasteners
  11. Pipe threads
  12. Metric threads

(After this block you will be able to identify thread types, determine uses, and use appropriate nomenclature relating to threads.)

P. Plastic Piping

  1. Introduction to plastic pipe
  2. Types of plastic pipe
  3. Chemical/material compatibility
  4. Plastic pipe schedules
  5. Fittings and connections

 6. Threading plastic pipe
 7. Miscellaneous accessories
 8. Pipe runs
 9. Pipe runs
10. Pipe runs
11. Pipe runs
12. Pipe runs

(You will learn pipe joining processes, chemical properties, physical characteristics, and types of accessories made of different grades of plastic.

Tailoring It to Your Needs

One thing about MCVTS that is particularly impressive is that this school is very "user-friendly"—that is, it really orients its facilities and activities to what the students and the community need and want.

When local industries or contractors have specific needs for training, they can go to this school, make these needs known, and the school will adapt its curricula to those needs.

Similarly, when students themselves see areas in which they would like more development, they can go to the school and mix and match their courses for specific objectives.

If, for example, the contractor who is your employer/sponsor in a plumbing apprenticeship program also does sheet metal work for a warm air heating business, and the two of you agree, you might be able to steer your courses toward that specialty.

The training overlaps for the first year. Then the sheet metal program goes into 36-hour blocks that include basic layout with sheet metal, pattern development and marking, stretch-outs of curved patterns, layout of duct run systems, forming round work, seams and edging, geometric layout, round geometry, triangulation theory, offsets and drops, pattern forming, and complex fittings.

While those course titles aren't fully informative as to what they are, they can give you at least a hint of what the work entails.

There are other job classifications that might be based largely on plumbing or pipe fitting but cover considerably more, and other blocks, or 36-hour modules, are offered for them.

For example, many skilled workers spend their careers in maintenance, which covers many trades but really requires full training and skill in one. The MCVTS course for "Industrial Maintenance Mechanic" is a four-year program with 16 modules of 36 hours each. Briefly, the 16 modules are the following:

1. Basic industrial math
2. Basic blueprint reading I
3. Math II for machine trades
4. Blueprint reading II, machine trades
5. Introduction to industrial wiring
6. Electromechanical processes
7. Principles of process piping
8. Mechanical processes
9. Welding theory I
10. Flat and horizontal welding
11. Welding theory II
12. Vertical welding techniques
13. Pneumatics I
14. Pneumatics II
15. Basic lathe operations
16. Milling

While a school will adapt as much as possible to your needs, don't make the mistake of thinking it can be frivolous.

The apprentice program is dead serious. You have to put in the hours or you don't make it. If in the four-year program you are out some nights because of sickness or other reasons, you must make those hours up. The only way to reduce the required 576 hours in the classroom is to pass an equivalency test. For example, if you have had college math you might be able to get credit for the math hours.

TYPES OF PROGRAMS

There are six arrangements, or forms of agreement, under which you might get into an apprenticeship program. All of them are based on a written program that details all of the provisions for both equipment and training.

Before you undertake this serious career move you should be able to see what the program is. In most respects it should parallel the program we already have outlined. Your own program might be any one of these six, and in some cases it will be more than one.

The six arrangements are the following:

1. A written agreement between you and your employer or the employer's designated agent. The agent might be a union, an employers association, or a joint apprenticeship committee.
2. A written program agreement between the employer or the employers association and the union describing the terms and conditions for employment and training of apprentices.
3. A written program plan prepared by the employer or employers association for firms without a union, that describes terms and conditions for employment and training of apprentices.
4. A written program prepared by the union that de-

scribes terms and conditions for employment and training of apprentices and has the employer's written consent.

5. A written program prepared by the employer or employers association that describes all these terms and conditions and has the union's written consent.

6. A collective bargaining agreement containing the basic standards of apprenticeship, with any supplements needed to cover these standards.

You might find yourself in a situation in which you know a plumbing contractor who wants you to work and for whom you want to work, but the contractor knows nothing about an apprentice plan.

That needn't be a problem. Under the National Apprenticeship Act, the BAT is responsible for providing assistance and service to this contractor (or any bona fide contractor or union or employer group) in setting up a program that meets the bureau's requirements. See appendix A for a list of state offices and their addresses and phone numbers.

Any such program gets formal recognition when it is registered with a state agency that is recognized by the BAT. If there should be no state agency, the program can be registered directly with the BAT. This registration is recommended for all programs. However, it is required when federal funds or benefits are involved.

And your contractor friend who hired you, incidentally, does not have to put up with some cut-and-dried program developed far away. It is the responsibility of BAT to customize the program to fit the needs of the particular employer.

COMPETITION

So far we have talked about "friendly" situations, but the apprenticeship program is not all roses. In many areas of the country the competition to get in a program is high, and in these places all prospective apprentices must apply for apprenticeship, undergo tests, and then go on a register. A register is a waiting list, and the wait can last for months, even years.

Openings for new apprentices in many cases occur only once or twice a year. For many contractors and other employers the needs can be seasonal, such as for those in construction work, and they might recruit only in the warmer months.

PROBLEMS FOR WOMEN

Women have special problems, and they probably already know about them. The fact is that plumbing, pipe fitting, and sprinkler fitting always have been a male domain. More women are doing it now, but many of them feel they are breaking into a man's world, so they need courage and self-confidence.

Regardless of their abilities they have to contend with the stereotyped attitudes of many of their male coworkers. Some men try to protect women workers from heavy or dirty work. But others try to make it much harder for women simply because "they don't belong here."

So the toughest part of the job for women is the breaking of the male taboo.

That, however, is why affirmative action is compulsory, and any employer with five or more employees must have a written affirmative action plan. That's the law.

APPLYING AND TESTING

We mentioned applying and testing. Let's look into that a little more.

There are programs that require applicants to pass certain tests to be sure they have good aptitudes for the work. For example, an applicant might be required to take a Specific Aptitude Test Battery (SATB) administered by a state job service agency. Such a test will measure two or more of the following nine general aptitudes: (1) general learning ability, (2) verbal aptitude, (3) numerical aptitude, (4) spatial aptitude, (5) form perception (the ability to perceive small details in an object), (6) clerical perception (the ability to distinguish pertinent detail), (7) motor coordination, (8) finger dexterity, and (9) manual dexterity.

Each battery tests different combinations of these aptitudes, because different occupations require different specific abilities. Some of these tests have written questions, and some use pegboards or other apparatuses to measure such abilities as dexterity. The scores are graded as *low, medium,* or *high.* All nine aptitudes are tested by the General Aptitude Test Battery (GATB) that is frequently used by counselors to help those interested in apprenticeship decide which trade to pursue.

Some tests are devised and administered by hiring companies. They often measure an applicant's familiarity with tools and terms of the trade.

If you get nervous or have other problems taking tests, you can get a booklet published by the United States Employment Service called *Doing Your Best on Aptitude Tests.* It might help.

"TASK-ORIENTED" ABC MODULES

Back in chapter 3 when we discussed the ways to get educated in this trade, we mentioned the task-oriented modules of the Associated Building Contractors and promised to show how they differ from the practical theory of other apprenticeship programs.

If you'll go back and hold your finger at the page where the MCVTS curriculum starts, you can compare the subjects with the following ABC subjects. These are presented only briefly here, by heading and with the number of hours you would spend on each.

Plumbing I (First year, 140 hours)

The first seven topics here are not basically different from those in the MCVTS curriculum:

- The plumbing trade, 2.5 hours
- Basic plumbing tools, 10 hours
- Basic plumbing safety, 5 hours
- Pipes, fittings, and basic terminology, 10 hours
- Math for plumbers, 15 hours
- Introduction to plumbing blueprint reading, 5 hours
- Reading residential plumbing drawings, 10 hours

But now comes a shift to specific tasks:

- Joining plastic pipe and fittings, 5 hours
- Soldering and brazing copper pipes and fittings, 10 hours
- Joining galvanized black pipe and fittings, 7.5 hours
- Joining no-hub and fittings, 5 hours
- Joining cast-iron hub and spigot pipe and fittings, 10 hours
- Making flare and compression joints with copper pipe, 5 hours

- Installing traps and interceptors, 10 hours
- Fitting and clean-out requirements for DWV piping, 15 hours
- Installing natural gas piping systems, 5 hours
- Installing liquefied petroleum gas (LPG) piping systems, 5 hours
- Installing fuel oil piping systems, 5 hours

Plumbing II (Second year, 150 hours)

- Reading light commercial plumbing drawings, 20 hours
- Installing pipe in trenches, 5 hours
- Grade for drain and waste piping, 7.5 hours
- Joining clay and concrete pipe, 2.5 hours
- Connecting to the sewer main, 5 hours
- Installing roof, floor, and area drains, 12.5 hours
- Installing pipe hangers and supports, 2.5 hours
- Installing DWV piping systems, 10 hours
- Testing DWV piping systems, 5 hours
- Connecting to the water main, 5 hours
- Testing water supply piping, 7.5 hours
- Types of faucets, 7.5 hours
- Types of valves, 5 hours
- Installing valves and faucets, 7.5 hours
- Servicing valves and faucets, 10 hours
- Installing water heaters, 5 hours
- Water meters, 2.5 hours
- Types of fixtures, 10 hours
- Setting fixtures, 15 hours
- Installing gasoline piping systems, 5 hours

A career in plumbing and pipe fitting requires reasonable physical strength and an ability to endure working in bad weather. (photo courtesy of Patrick Galvin)

Plumbing III (Third year, 150 hours)

- Reading commercial plumbing drawings, 5 hours
- Introduction to regional and local plumbing codes, 5 hours
- Types of vents and installation specifications, 15 hours
- Indirect and special waste, 7.5 hours
- Sewage pumps and sump pumps, 10 hours
- Locating buried sewer and water lines, 5 hours
- Materials for storm drainage systems, 2.5 hours
- Installing water pressure reducing valves, 2.5 hours
- Water pressure booster systems, 7.5 hours
- Installing shock arrestors, 5 hours
- Backflow preventers, 5 hours
- Installing recirculation systems, 5 hours
- Filtering and softening water, 5 hours
- Cleaning and disinfecting potable water systems, 5 hours
- Thawing frozen water pipes, 2.5 hours
- Installing water supply piping, 10 hours
- Fixture rough-in, 7.5 hours
- Plumbing for solar systems, 5 hours
- Piping and venting natural gas systems, 5 hours
- Installing private waste disposal systems, 10 hours

Plumbing IV (Fourth year, 150 hours)

- Plumbing theory, 15 hours
- On-the-job task organization, 7.5 hours
- Sizing storm drainage, 5 hours
- Sizing drainage systems, 10 hours
- Combination waste and vent systems, 7.5 hours
- Servicing traps and interceptors, 5 hours
- Sizing water supply piping, 10 hours

- Servicing plumbing fixtures, 15 hours
- Installing fire protection systems, 17.5 hours
- Installing swimming pools and hot tubs, 5 hours
- Installing compressed air piping systems, 7.5 hours
- Installing hydronic heating systems, 7.5 hours
- Installing corrosive-resistant waste piping, 7.5 hours
- Installing medical gas systems, 7.5 hours
- Plumbing for mobile homes and mobile home parks, 7.5 hours
- Installing private water supply systems, 10 hours

As pointed out previously, you can switch over to get other specific modules in pipe fitting or HVAC according to the needs of your job. You can do the same at MCVTS, except that it has to be done with the approval and cooperation of the committee that administers the program.

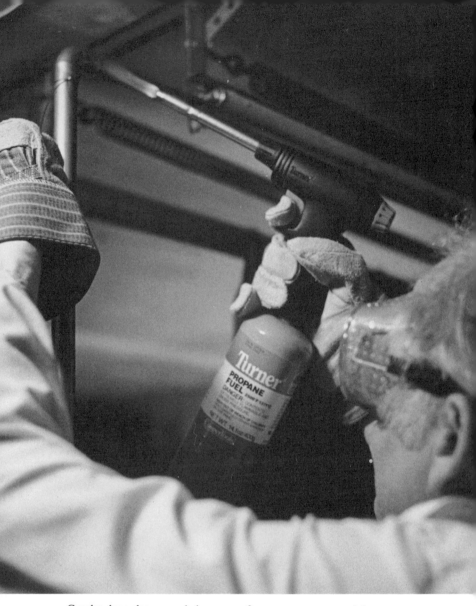

Goggles, long sleeves, and gloves are often a necessary part of the plumber's uniform. (photo by Bob Wendlinger)

CHAPTER 5

A TYPICAL WORKDAY

Plumbers usually start early. The aim generally is for an eight-hour workday, but emergency is part of the nature of the work. Water pipes can freeze, burst, or spring leaks at any hour of day or night. Sewer lines don't care what time it is when they get stopped up.

That's bad? No, that's good. That's what makes plumbers so necessary at all hours of day or night. When people really need you, it is easy to find work and get good pay.

But that's the day of the person we usually visualize when we think of a plumber.

There are other plumbers who work for firms that do commercial jobs, and they might have more regular hours. There are plumbers who work for governmental units—schools, counties, cities, states, and even the federal government— and they usually don't worry so much about random emergencies, because their responsibilities usually are at one building or group of buildings.

There are others who work on big jobs as members of large crews—building convention centers, banks, office buildings, shopping centers, hotels, apartment complexes, or new home developments—and they seldom have the problem of emergencies.

So pick the scenario you like. Once you get through the apprentice period, the choice is yours.

Here, let's look at some of the scenarios.

THE "REGULAR" PLUMBER'S DAY

This is the person we mentioned first.

6 A.M. Up at the crack of dawn. (Well, we just said that because it sounds good. Actually, we never heard dawn crack, and besides, if you are in New York dawn was a half-hour ago, and if you're in San Antonio it might not come for another hour.) But you have to be up and ready to run.

7 A.M. Out the door. Maybe you are heading for breakfast, or maybe you've eaten and are heading for the shop. In any case it is quite likely that you won't be driving your own car. Instead, you probably have a company truck at home, and it has two-way radio. So you push a button and report that you're on the way.

7:05 A.M. The office reports back to you in your truck: "Good morning, Joe. Hey, we got a call from a Mrs. Jones at 4444 Buena Vista. She's got no hot water." You answer: "OK, that's only about a half-mile from here. I can catch it on the way in." The office says to go ahead and do it.

7:17 A.M. You check in at 4444 Buena Vista and report to the office that you have arrived on the job. Mrs. Jones says, "My gosh, I just hung up the phone. How did you get here so fast?" You answer, "It's all in the day's work, ma'am. When somebody has a problem we'll always make it as fast as we can." As you head for the basement you notice three children, all preschoolers, eating breakfast. You go down and check the water heater. No problem. It is a gas water heater, and the pilot went out.

From this point on there are two ways it can go, depending

partly on the kind of plumber you are and partly on the kind of firm you work for. It can go (1) the routine way or (2) the creative way.

1. The routine way. You relight the pilot and check it and the piping around it to make sure everything is all right. Then you report back to the woman who called: "It's all right now, ma'am, but it will take a couple of hours to recover. If you really need hot water soon, you'll have to heat it on the stove. That's $35, our minimum for the call. You can give me a check, or we take credit cards."

7:30 A.M. Back in the truck you pick up the radio: "This is Joe. I got the Buena Vista call. It was just a pilot. I'll be there in about 25 minutes."

2. The creative way. You relight the pilot and check the piping around. Then you think. You know the houses in this development were built in 1970, and this 40-gallon water heater is original equipment. That means it is 18 years old. You know that the average life of a water heater is about 15 years. You noticed three preschoolers upstairs, and that means the demand for hot water has increased. That tank now has to satisfy the needs of five people instead of the two who moved in 18 years ago. Besides, water heaters are made better now. They are insulated better to save energy. The cost of energy was insignificant 18 years ago, but it is very significant now.

So you go back upstairs and tell the woman it is fixed and she owes you $35, but you also point out to her that the old water heater has just about run its course and might be due for replacement at any time. And since the family has grown, perhaps that old water heater is too small.

You explain that a 40-gallon gas water heater will provide about 73.6 gallons of hot water in the first hour, and that sounds like a lot, but a family of five can use 50 to 70 gallons of water just in taking five showers in the morning—even if they are conservation conscious. If two of those people each take five-minute showers, they alone will use 60 gallons. And that doesn't allow for doing any laundry or running hot water in the kitchen.

You point out to her that life can be pretty miserable if the water heater conks out on a cold winter morning, especially if service isn't available quickly. And you suggest that she consider getting a new, larger water heater now, before the inevitable happens. And you ask her: "Shall I ask the boss to call you about it?"

7:40 A.M. Back in your truck, you call in: "This is Joe. I took care of it on Buena Vista. It was just a pilot. But these people need a new and larger water heater. I explained it to the woman, and she said they'd talk it over and you can call her in a couple of days. I'll be there in about 25 minutes."

(What do you get from this extra effort? First, you make yourself more valuable on the job. Second, you add a little to the prosperity of the firm you work for. Third, you get a commission on the sale of new equipment.)

8:10 A.M. You check in at the shop. The dispatcher has several calls for service, but you are scheduled to install an ice-maker in the two-year-old refrigerator of an old customer. You note on the job ticket that this is a model made to accept later installation of an ice-maker, and the customer already has bought the ice-maker, so basically it is a plumbing job. You know nothing about the layout in the house, so you go to the supply room and draw out a 50-foot coil of 1/4-inch polyethylene tubing, a couple of self-tapping slip compression saddle valves, and two brass inserts to reinforce the ends of the tubing. You check your truck to be sure you have copper tubing and a compression union for connecting your new tubing to the connection on the back of the refrigerator.

9:20 A.M. You arrive at the customer's house, report to the shop on your radio, introduce yourself, and inspect the new ice-maker and the installation directions that were included with it.

It is clear enough and simple enough. Your problem is to find a cold water supply line to which you can attach the tubing, and then get it to the back of the refrigerator where it will connect with a fitting already built into the refrigerator.

The water supply lines at the sink are across the kitchen from the refrigerator. You can get water from there. But you notice there is a half-bath across the hall from the kitchen, so you know there probably will be a cold water supply line going to the half-bath. You go down to the basement to check the pipes and sure enough, at the basement ceiling there is a 3/4-inch line of copper tubing going from the kitchen sink supply across to the half-bath, and you can tap into it at a point only about six feet from where you can bring it up through the floor behind the refrigerator. You check for other possibilities, but this is the best choice.

You notify the house occupant that you will have to turn the water off for about an hour. You turn off the house supply and ask the occupant to open all faucets in the house. This will avoid a puddle in the basement.

You attach your self-tapping valve to the tubing with clamps that come with it, turn the handle on the valve, and it taps into the line. You now can make a choice of using the plastic tubing or the copper tubing you brought with you. If access is difficult and there must be a lot of bends and turns, the plastic would be better. But this is a clean, straight run, and being a pro at your job, you know the copper makes a more permanent installation. So you opt for copper. You then attach a line of 1/4-inch copper tubing to the new compression valve and run it over to a spot at the wall just under the refrigerator, drill a hole there, and bring it up through the floor.

Following the instructions that came with the ice-maker, you hook everything up properly.

This job has not been difficult, but it has been time-consuming. And wanting to get it done, you stayed on the job right through your lunch period.

1:05 P.M. You call in on the radio: "I finished the ice-maker job, and now I'm going to take a lunch break." The reply comes back: "OK, Joe. Come to the shop when you finish."

2:10 P.M. You arrive back at the shop. Since you started early on this day with the hot water emergency call, your day is scheduled to end at 3 P.M. If there should be another emergency you would go out on it, if needed, but things are well in hand. The job expediter tells you that tomorrow you will rough-in a new bathroom in an older house, and you will be expected there at 8 A.M. You use your remaining hour of work on this day to check the stock in your truck for the materials and tools you will need for the rough-in, filling in all that might be needed. At 3 P.M. you head for home.

"NEW CONSTRUCTION" WORK

There are other "regular plumbers" whose day is not like that at all. They might be working on what the trade calls "new construction," which means a new development of speculative townhouses or detached houses.

These plumbers might go directly to the job each morning, begin at 8 A.M., and spend the day hooking up the kitchen appliances, bathroom fixtures, and water heaters or installing heating systems, knowing they are expected to complete a certain number of units each day. The equipment, fixtures, and materials will have been delivered to the job so they will be ready when these plumbers arrive.

Or possibly the plumbers will drive their own cars to the

shop, pick up a truck that has been outfitted with all the equipment they will need, and then go to the job site.

In either case, these journeymen and apprentices will be fairly free of supervision during the workday as long as they get their work done and are reasonably productive.

If they need more materials in the field, they often are authorized to go to a local plumbing and heating wholesaler to get what they need. But if this happens a lot, it will mean either that the trucks are not being stocked properly at the shop or that the crew is ruining too many materials on the job, and then the boss will want to know why.

Often these plumbers will spend their days in construction sites where they have to climb over piles of mud and refuse to get into the shells of houses where they will work. There will be no heat in the winter and no air-conditioning in the summer.

Usually the houses will be framed, and the roofs will be in place for most of their work. The local utilities will normally be responsible for getting supply lines to the houses, but such housing developments often are farther out from the city or town than utilities work, so the new construction plumber might be laying in those supply lines out in the open, in rain or snow or hot sun.

But usually they will work regular hours.

COMMERCIAL PLUMBERS

Commercial plumbers might work for the same contractors as the plumbers we have just described. But that is not always the case, because there are some plumbing firms that specialize in commercial work and do nothing in the residential field.

In many cases they will work a regular route, calling on customer firms and doing scheduled maintenance work. The idea here is to keep all plumbing, heating, and cooling equipment in good shape so potential problems are headed off before they happen. This is a nice, regular job with regular hours, normally.

But of course with any mechanical equipment strange things can happen, however well it is maintained. So if a department store's heating plant fails in the middle of a winter cold spell, or if its cooling plant conks out in mid-July, this is very serious for the store's business, and therefore it is critical for the plumbing firm responsible for maintenance.

Look at it this way. If you are a residential plumber and somebody calls in with an emergency problem and you can't respond to the call quickly and solve the problem, you are not really losing business. You simply are failing to gain some new business.

But in commercial work, the customer who calls is part of your regular customer list and is depending on you totally. If you can't take care of the problem immediately, you are losing business that might be critical to the plumbing firm's existence. There is no margin for error here. Whether it is noon or midnight, Saturday night or Christmas morning, the problem has to be attended to, and right away.

Of course, such dire emergencies won't happen often. If they do, your maintenance work is not good enough. Generally, it is a pleasant way to spend a working life.

WORKING FOR BIG-JOB CONTRACTORS

Life can be exciting and filled with variety and challenge when you are a journeyman working for a big-job contractor.

These are plumbing-heating-cooling firms, but because they do all of that plus process piping and more on such a big scale, they usually are called by a more comprehensive name—mechanical contractors. One major contract for one of these firms might be more than the annual volume of a dozen average-size firms, and a mechanical contractor with headquarters in, say, Ponca City, Oklahoma, might have contracts in Dallas, London, Singapore—and even Ponca City, Oklahoma.

So you, as a plumber working for one of these large firms, might find yourself closing out a job in Ponca City today and leaving for Singapore tomorrow. On any big job you will be on a regular work schedule, and you will know where you will be working not just tomorrow but for several weeks, even months, before shifting to another phase of the job.

Let's consider a representative big job, an actual new bank building in Texas.

This will soar 600 feet above the street, with an eight-story bank wing and a 36-story tower, plus four subterranean floors devoted mostly to drive-in banking and parking, and two more levels below for the "mechanical equipment," the nerve centers for cooling and heating. On the lowest level, 80 feet below the street, is the refrigeration room, with two 1,000-ton centrifugal refrigeration machines. Up one level are two gas-fired water-tube boilers each rated at 25,000 pounds per hour to drive the two 1,000-horsepower steam turbines that drive the refrigeration machines.

This job will take 500,000 man-hours of labor over four years, and in that time you will be working with as many as 195 other plumbers, pipe fitters, steam fitters, and sheet metal and refrigeration specialists at one time. Heating and cooling must be provided for a total of 10.6 million cubic feet of space. It will take 49 separate air-conditioning systems for combination heating and cooling and eight additional units for heating only such places as the garage entrances and exits and other doorways to the building.

The major air-conditioning breaks down into three basic systems: (1) a high-pressure conduit system for the periphery zone of the tower section, which extends all the way down to the street; (2) interior zone units for each floor of the tower section except the seventh, and excluding the first four floors that are below street level; (3) and 15 conventional dehumidifier systems to condition the air in the space below the seventh floor plus the auditorium on the eighth floor, and two more to serve the periphery system in the tower.

All 17 of these dehumidifier systems are on the seventh floor, which is devoted entirely to machinery. There are two more conventional dehumidifier systems on the 36th floor, which is another floor given over entirely to machinery. These systems also are for the periphery zone. Conditioned hot or cool air will be delivered to 2,543 units, which will give individual control throughout the bank by adjustment at the unit of the hot or chilled water flow.

Four 150-horsepower pumps handle primary chilled water and condenser water. Secondary chilled water is handled by four other pumps, two of them 40 horsepower and two 25 horsepower. From this location, 80 feet below street level, it is more than 500 feet up to the big cooling tower based on the 36th floor and rising to the top of the building. The cooling tower, matched to 2,000 tons of refrigeration, is rated at 6,600 gallons per minute.

Back downstairs to the fifth sublevel, the huge boilers must generate more steam to cool the building than to heat it. In winter they need to produce only enough heat to warm the building, which is not a big deal in Dallas. But in summer they must produce 30,000 pounds of steam to drive the turbines for the cooling load, which requires 600,000 cubic feet per minute of refrigerated air. Rising from the boilers is the city's tallest smokestack, but you won't see it on the Dallas skyline because it is all inside the building.

The plumbing part of this contract had to provide for what was unknowable—where tenants would want private bathrooms. Plans called for restrooms for women and men on each floor of the tower section and at various points in the bank section. To anticipate the desires and needs of tenants, the contractor installed "wet" columns running up through the tower section so complete plumbing would be available on demand simply by tapping into these wet columns. Normal, preplanned plumbing worked off specified risers. A 24,000-gallon, two-compartment house tank is on the 36th floor and is supplied by two 50-horsepower pumps in the boiler room.

Piping sizes ranged from 1/8-inch copper tubing to 20-inch pipe for the turbine exhaust. The job took 1,500,000 pounds of sheet metal, which was fabricated into ductwork and installed by this same big-job contractor.

Because of the height of the building, all valves and fittings from the 15th floor down had to be extra heavy to withstand the pressure. For example, 300-pound valves had to be used below the 15th floor because of the static head.

There are all kinds of aggravating problems on a job such as this. For example, where will you put all of this material before you install it? How will you get 24,800 pounds of house tanks up to the 36th floor, especially when there is no 36th floor—just a lot of steel girders?

These are problems that have to be solved as they come up. And there is nothing easy about it, especially in a building under construction. But the question of the house tanks required a bit of cooperation with another big contractor whose business was steel erection. The other contractor hoisted the tanks up to the 36th floor and lashed them to the girders where they waited until the floor was in. In return, the big-job mechanical contractor did a lot of hoisting and moving for the steel contractor.

That is a rather long story about a typical big job. And you might well ask what it has to do with a young person just now contemplating a career in plumbing and pipe fitting.

Plenty. Because it is a story of challenge. It is not a matter of going to the job every day for four years and doing the job according to the blueprints.

Because like every big job, the rules keep changing every day. Parts of the building must be completed and ready for occupancy every day, even though the rest of the building still is under construction. That means changes, constant changes. There are deadlines in every phase of the job, and the deadlines keep changing day by day. That means new plans have to be developed as the building progresses.

On this job, seven engineers worked full time on coordinated drawings, and every day the work had to be coordinated with the other building trades.

It is a fact that it was two years after the building started before our big-job contractor had a complete set of plans; and they added up to 3½ tons of blueprints, specifications, and other paper.

And with tenants making constant changes in design, it was always necessary that plumbing, heating, and cooling be available in completed parts of the building while other parts were under construction. And this had to be done always with the idea in mind that when the bank shut down for bank holidays there would be no effect on tenants who remained open for business.

In other words, despite all the executives and highly paid engineers, there always was a plumber with a wrench in hand who had to make a decision.

And while we indicated at the start of this story that working for a big-job contractor meant fairly regular hours, it is not necessarily true all of the time. Even in big-job contracting there can be a series of emergencies. Actually, that is not accurate. It is more accurate to say there always will be a series of emergencies. The sheet metal won't be delivered on time. Or the boiler factory will go on strike and be held up for three months. Or there will be a three-day snow storm, or rain storm, and the materials or fixtures can't be hoisted to the right floor at the right time. Maybe the hoist itself breaks down.

It can get even worse. Remember, in a high rise, everything has to go up on construction elevators, and they have to go on schedule. If you miss the schedule for taking up 90 toilets and 90 bathtubs, it is possible that you won't be able to get another hoist day for two or three weeks, because the construction elevators are scheduled to hoist lumber or steel or flooring for some other subcontractor and all these materials are on time.

That can mean a lot of overtime work, but it also can tax your ingenuity to keep working on a productive basis even though the scheduled material or equipment is missing.

The job always is to solve problems, whether you are working for a big-job contractor or a neighborhood plumber. There are as many kinds of challenges as there are plumbers working on the many different kinds of jobs. And for many, that is what makes it worth doing.

One career option for plumbers is to work for a wholesaler as a remodeling manager. (*Plumbing & Mechanical* magazine photo)

CHAPTER 6

YOUR FUTURE IN THIS FIELD

Let's just dream for a little while. Suppose you put in all that effort, all that classroom time at night, and all those 8,000 hours of on-the-job work. Is there any chance that you can really make a bundle?

There's only one answer, and that is: of course you can.

But that answer has to be modified. It depends on the kind of person you are.

First, consider it realistically. You have known people all your life who drove the biggest cars and lived in the nicest houses and went to Hawaii every winter—plumbers, grocers, store owners, electricians, bankers, janitors, cops. And you have known others in the same field who always were on the verge of—or in—bankruptcy.

What makes the difference?

Three things. First, you have to have ideas.

Second, you have to have them in a field where you have some expertise.

And third, you have to implement them. You have to have the guts and stamina to do something with them.

But wait, you might say. You are only considering starting a plumbing apprentice program. All you want is a job, a way to make a living.

Fine. That's the way it starts. That's the way it has to start.

But you don't know who you will be 20 years from now. Quite probably, you don't even know who you are now.

And that is understandable. You don't yet have a trade, or at least you don't have one in which you are comfortable. You don't have anywhere near the skills you will have four years from now when you finish an apprenticeship. Quite possibly you don't even have a high school diploma yet.

So how can you know now who you are or what you can be?

Let's look ahead at what the future holds for you.

A RUNNING START

First, you are looking at four or five years of hard work and hard study. In this period you'll be paid for the work. It won't be a lot, but it should be adequate. As a starting apprentice you will be paid 40 or 50 percent of journeyman pay, and this will increase every six months.

Then you'll go out into the world as a journeyman, skilled in a trade. And this is something you will take with you everywhere.

Where might you work?

You already have read about what a day can be like in the life of a plumber or a pipe fitter. Any way you cut it, it is a good life with good pay. And that might be fine with you.

But remember. You don't know who you are yet. Whoever and whatever you are right now, you will be a different person after four or five years of dedication to a single purpose, the purpose of completing the training and becoming a journeyman in a specific trade.

GROWTH IN THE TRADE

Here's a scenario of how your job will progress.

For the first four or five years you will work on the job with your sponsoring contractor and will go to school two nights a week. In this period your pay will increase to full journeyman scale.

Then you have full freedom to go where you want and where you can get the job. The next step, whether you stay where you are or move, is to become a foreman. That means higher pay, which might be arranged between you and your employer or might be according to a union contract.

Moving into Management

From that point on you can move into management. That can mean anything, depending on the kind of firm you are with.

You might be with a big-job contractor, the kind that gets plumbing and fitting contracts to install all of the complex plumbing and piping and ductwork in a huge new bank building or apartment complex, or in a stadium or chemical plant. This contractor generally will be known as a mechanical contractor. In this situation you can go on to become a superintendent or a project manager or to some other title, but it will definitely mean moving into a high-pay bracket.

Opportunity with Small Firms

But you also can move up if you stay with a smaller plumbing contractor. The smaller firm might have 5 or 50 plumbers working. *Smaller* here refers to the size of the firm's contracts rather than the number of people in the firm.

You can become a manager of the entire plumbing force. Or you might become a manager of the dozen or so who work on specific types of jobs, such as installation of kitchens and bathrooms in the remodeling department or in the new home department.

It was in the middle 1950s that kitchen remodeling became an industry unto itself. In the middle 1960s bathrooms followed. Generally, it was the plumbing contractors who opened new divisions, or departments, to work in the remodeling field.

The Need for Bosses

Such a new division needs managers, and that is where you come in, because this still is happening today. Where plumbing contractors are not doing kitchen and bathroom remodeling, plumbing wholesalers are stepping in to fill this need, and they need people like you.

We haven't mentioned the wholesaler before. Wholesalers are important links in the distribution chain that moves products from manufacturer to market.

Wholesalers buy from manufacturers and sell to plumbing, heating, and HVAC contractors. They provide the warehousing and yards where all the many types of pipe, fittings, bath fixtures, and other products are kept until the contractors need them. The contractors themselves often don't have such facilities or the large amounts of money to buy by the carload. Manufacturers want to sell in these large quantities. They have neither the time nor the personnel to sell products one by one to a plumbing contractor.

Why Wholesalers Need You

Wholesalers also do not have the personnel or the contracting expertise to operate a kitchen and bathroom remodeling business.

So the first thing they need, in the planning stage, is a remodeling manager. This manager might be an accomplished salesperson, an accomplished designer, an accomplished business manager, or an accomplished plumber.

In very many cases, probably in most cases, wholesalers make mistakes in hiring their first remodeling manager, because they hire a sales or design person who has no business sense or a business manager who has no plumbing sense.

Where You Come In

And that should suggest things for an apprentice to think about right from the start of training.

First, of course, you think about your work and your study. But any worker in this field can prepare for a better future by also always being aware of how people sell and how people run the business. Information on how to design kitchens and bathrooms can come from reading any of the many books available at any bookstore, home center, or library. Or for the real pros there are many schools available. There are both basic and advanced schools on kitchen and bathroom design put on regionally by the National Kitchen and Bath Association, and several retired kitchen and bath pros are available nationwide for training purposes.

So that can be one of opportunity's knocks. But there can be a better knock, even, than that.

HAVING YOUR OWN BUSINESS

The best way, if you have the ambition and the willingness to work hard and if you are suited to it, is eventually to go into business for yourself.

The plumbing field is excellent for that for several reasons.

One reason is that the opportunity is there in many areas of the country. Many workers prefer to work for others. They don't want the hassle of running a business and figuring out a payroll. They don't want the problems of filling out all the necessary government forms and paying other people's taxes. So there are not a lot of new plumbing businesses starting.

Another reason is that many of the plumbing businesses are many years old and are not quite up to modern merchandising and advertising. The starting apprentice, simply by being younger, might be more aware of merchandising and advertising strategies.

A third reason is that there nearly always is a need in any community for another skilled plumbing business.

The fourth reason is that there are so many new opportunities in remodeling, and this is true of most areas of the country. In almost any city or town, a new plumber can set up as a kitchen and bathroom remodeling specialist. If he or she doesn't feel comfortable with running such a business, then he or she can set up as an installation contractor for kitchens and bathrooms, concentrating just on installation. And this person will find many eager customers among the kitchen and bath remodelers who do not do their own installation and don't want to.

EXCITING NEW FIELDS

But there are exciting new opportunities developing every day in other new fields as well. We have mentioned them already, but consider these possibilities:

1. *Space.* Plumbers will be as necessary as astronauts in our future in space, and there will have to be a lot more plumbers than pilots. It is a simple fact that where there is a need for water, heating, cooling, venting, and waste disposal, the plumber must be there.

In the future, many new jobs will exist for plumbers in the petrochemical industry. (photo courtesy of the American Petroleum Institute Photo Library)

2. *Geothermal energy.* The natural heat deep in the earth soars to 2,000° F when we just dig down a mile. We see evidence of it in the boiling lava that flows from volcanoes daily. But more important, there already are geothermal energy plants in existence in the United States that use this heat to generate electricity. Such plants need piping, a lot of it.

3. *Solar energy.* This was big news in the 1970s. The only reasons for a loss of interest were a drop in oil prices and the need to wait for new technology to catch up. Technology is catching up right now, and nobody knows about oil prices in the future. Solar energy is in the cards for the years ahead, and every plant will need piping—lots of it.

4. *Petrochemicals.* Wherever you see plastics you see petrochemical products. They are made from oil or gas. One big petrochemical plant can need hundreds of miles of piping, even thousands of miles. This is another area that means many, many new jobs in future years in an exciting new industry.

5. *Oil.* Production has been diminishing in recent years, but only because worldwide prices are so low that it is cheaper to buy it abroad than to produce it here. But it is in the ground in the United States, and sooner or later the wells will open up again. Then there will be more need for more refineries that, like petrochemical plants, require hundreds or thousands of miles of piping.

6. *Desalinated water.* Fresh water from the oceans has been one of humanity's dreams for generations. It is a reality now, although the technology is young and therefore not widespread. But this definitely is one of the fields of the future for plumbers and fitters.

MULTIPLYING OPPORTUNITIES

So obviously it is not enough to say the plumbing and fitting field is growing. It is mushrooming, and that means the opportunities for today's beginners are multiplying. Not only are job opportunities multiplying, they also are multiplying in specialty areas that mean higher pay.

That is true because specialty areas, being new, always offer better pay, because they have to attract the labor force away from what it already is used to.

That's a lot to think about for someone who is just now considering becoming an apprentice in plumbing or pipe, steam, or sprinkler fitting. But it is all there waiting for the work force of tomorrow.

Air-conditioning and refrigeration mechanics install and service central air-conditioning systems and refrigeration equipment. (Lenox Industries, Inc. photo)

CHAPTER 7

OPPORTUNITIES IN RELATED FIELDS

We already have mentioned some fields related to plumbing and pipe fitting. At some time in your apprenticeship program you might decide you are more interested in one of them, so let's take a closer look at them now.

SHEET METAL WORK

Sheet metal workers usually work in the field of warm air heating. This might be residential or commercial.

They make, install, and maintain a wide variety of products, including ducts for heating, cooling, ventilating, and pollution control systems, kitchen equipment, roofs, siding, rain gutters, skylights, and other applications. You can see this shiny metal ductwork in almost any residential basement.

Most sheet metal workers in the construction industry work in all three jobs: making the ducts, installing them, and maintaining or repairing them. Their raw material comes in flat sheets, at times presized and precut in specific lengths and widths, and the worker forms these pieces into runs of ductwork. The worker forms elbows, tees, and other turns and fittings as needed and then joins them all together.

More commonly these products are formed in the shop of the sheet metal or heating contractor. There the sheet metal worker works from blueprints and instructions from supervisors to measure, cut, bend, and shape, but much of the joining must be done at the job site.

In many shops the metalworking equipment is computerized for layout to get the most usage and least waste from large, flat sheets. Saws, shears, and presses are computerized. In some shops cutting is done with computer-controlled lasers.

Where computers are not used, either because the shops don't have them or because the jobs are too customized for mass production techniques, layouts must be done with tapes, rulers, hand-held calculators, and other measuring devices, and then the parts are cut or stamped out with machine tools.

As much as possible is joined in the shop by bolts, cement, drive slips, rivets, or solder, or by welding. At the job site all of the prefabricated parts are assembled.

There might be quite a bit of alteration needed at the job site because of errors or inaccuracies in the blueprints or because the blueprints were not followed precisely by other construction trades. It is not unusual for walls or doors to be misplaced by a few inches during construction, and this can make a duct run impossible. Another factor is late change. Plans and blueprints often are changed during construction, and this can alter the duct runs.

The ductwork is installed as it is assembled. Metal hangers are used to hang it from ceilings or to secure it to walls. It often has to be fitted in between floor joists of the floor above and between walls of adjacent rooms.

In the process of assembling on the job, joining is done with bolts, welding, rivets, glue, or solder and with the specially formed drive slips that are cut and shaped into the parts at the shop to insure fast installation. This is necessary, because job site labor can be more difficult, and usually costs more, than shop labor.

Roofing and Siding

When sheet metal is used for roofing and siding, measuring and cutting usually are done on the job.

After the first panel is secured in place, workers interlock and fasten the molded edge of the next panel into the formed edge of the first. The free edge of the panel is nailed or welded to the structure. At joints, along corners, and around windows and doors, workers fasten a machine-made molding to finish the job.

Hard and Heavy Work

This is not easy work. Metal obviously is heavy, and a lot of the lifting and moving has to be done squatting in close quarters and in awkward positions. Sheet metal workers get a lot of cuts and minor burns from the materials and tools.

However, according to government statistics, sheet metal workers lose less work time due to bad weather than other construction workers. Except for roofing and siding, more work is done indoors than in the other trades.

Jobs and Earnings

Generally, there are only about one-fourth as many jobs for sheet metal workers as there are for plumbers. Pay can be considerably less. For example, in a year when median pay for a plumber was $405 per week, the median pay for a sheet metal worker also was $405, but whereas the top 10 percent of plumbers earned $741 per week, the top 10 percent of sheet metal workers earned $680. For roofers, the median in the same year was $285 a week and the top 10 percent earned $620.

Training and Apprenticeship

Apprenticeship is recommended for learning this trade, although many more learn it on the job informally than in plumbing.

Much of the apprentice program parallels plumbing and pipe fitting, but there also are many differences. You have to learn the skills of measuring, cutting, bending, fabricating, and installing, but then you have to get into complex fittings and decorative pieces. Near the end of training you also will learn the use of such materials as pressed fiberglass, plastics, and acoustical tile, which may be substituted for metal on some jobs.

In the classroom you have to learn drafting, blueprint reading, enough trigonometry and geometry to enable you to lay out the work, and the use of computerized equipment.

Going on Your Own

While it is fairly easy for a plumber to go out on his or her own as a contractor, this is much more difficult for a sheet metal worker. The reason is that it is much more expensive to set up shop. Modern equipment is expensive, and without modern computerized equipment it is difficult to produce at low cost and bid against established shops that have it. Good equipment lowers labor cost, and nearly all jobs in construction are won on a lowest-bid basis.

HEATING, COOLING, AND REFRIGERATION

Heating and air-conditioning systems control the temperature, humidity, and cleanliness of air in homes, schools, factories, and offices. Refrigeration systems sometimes are used for that cooling. They also make it possible to store foods, drugs, and other perishable items safely. The skilled workers who install, maintain, and repair these systems are called heating, air-conditioning, and refrigeration mechanics.

Several Appliances, Many Skills

Workers in this field get involved in several different types of equipment. For example, there are oil-fueled, electric-fueled, and gas-fueled furnaces. Many of these have add-on cooling units that involve compressors, condensors, refrigerants, and controls. There are vent pumps, relays, and thermostats.

Mechanics have to be skilled in all of these, and that includes plumbing, sheet metal, and wiring. In large construction jobs all of these different jobs are usually unionized, and that means only sheet metal workers can do the sheet metal work and only electricians can do the wiring. But in very many of the smaller jobs, each worker must be able to do it all.

What You Might Do

You might specialize in installation or in maintenance and service—or you might do it all.

If you are a heating equipment installer you might follow blueprints or drawings to install oil, gas, solid fuel, or multi-fuel heating systems. The equipment might be a furnace, which heats the air, or a boiler that heats water for either hot water or steam heat.

After setting the equipment in place you install fuel and water supply lines, air ducts and vents, pumps, and other components. Then you connect electrical wiring and controls and check for proper operation.

The system might even involve solar heating equipment, in which case you might have to install roof or yard panels to collect solar energy and then install a distribution system to distribute the heated air or water. The job might involve installing and connecting photovoltaic cells, which generate electricity from the sun.

Cooling and Refrigeration

Air-conditioning and refrigeration mechanics install and service central air-conditioning systems and a lot of different refrigeration equipment.

These mechanics install motors, compressors, condensing units, evaporators, and other components. They have to connect this equipment to ductwork or to water lines, to refrigerant lines, and to the power source. After making the connections they charge the system with refrigerant and get it going to make sure it works.

Working Conditions

Much of the work is in buildings that won't have any heat until you put it in. A lot of the work is in awkward or cramped positions, and you can be subject to muscle strains from lifting heavy equipment and to burns and electrical shock. Much of the work might also be in high places. You might, for example, have to install cooling towers on top of high-rise buildings. To some people, even a house roof might seem to be a high place.

Training and Jobs

Training in air-conditioning and refrigeration is available in four-year apprenticeship programs, although many workers start as helpers to mechanics and gain experience over the years.

Training includes microelectronics because of the increasing use of this technology in controls. After reaching journeyman status, many mechanics must continue training at schools conducted by equipment manufacturers and by the associations in the industry. This is because the technology changes and the equipment grows more complex.

Jobs are about twice as plentiful as in sheet metal work, about 40 percent as plentiful as for plumbers. However, jobs here are not as sensitive to economic downturns, because all people need heating, cooling, and refrigeration. So maintenance must go on, and there always is a need for installation of newer, more efficient equipment.

The job outlook is especially good in the South because of industrial growth there and because of many new households being established there, as many people from the North flee winter.

Median earnings of $370 per week in 1984 were about 10 percent less than for plumbers. The highest 10 percent earned $615 weekly, which compares with $741 for a plumber.

Mechanics who work on both heating and air-conditioning frequently get higher rates of pay than those who work on only one or the other.

SPRINKLER FITTING

Sprinkler fitting actually is a specialty within plumbing or pipe fitting. But the differences are sufficient to warrant separate treatment here.

Sprinkler fitters are pipe fitters who specialize in fire protection. They install automatic fire protection systems in homes, offices, hotels, factories, schools—just about any kind of structure.

The job requires persons who can carry heavy pieces of pipe of various lengths, climb and work at various heights, and handle the tools of the trade.

The sprinkler fitter installs, maintains, and repairs all types of fixed piping fire protection systems.

The Desire for Sprinklers

While there are many buildings that do not have such systems, everybody wants them.

In a typical year in the United States more than 6,000 people die as a result of fire and more than $7 billion in property is lost. And 71 percent of businesses that suffer severe fire loss close permanently within three years.

But there has been no multiple loss of life in fully sprinklered buildings, except for "human torch" fires and "cold" fires that involve smoke and poisonous gases only. Sprinkler systems perform satisfactorily 97 percent of the time, and in six out of ten cases where they operate they extinguish the fire without human aid.

The Need for Good Apprentices

The automatic sprinkler industry has a record of excellence, and the industry believes that excellence will depend in the future on the type of apprentices who enter the field. According to the four minimum standards the industry has established, an apprentice (1) must be at least 18 and have a birth certificate available at time of application, (2) must be a high school graduate or equivalent and show a completion certificate or diploma at time of application, (3) must be

A serviceman checks a new safety device on a furnace. (York Division,
Borg-Warner photo)

physically fit and undergo a physical examination to verify it, and (4) will be subject to an aptitude test and an oral interview.

Much of the apprentice training parallels that of plumbing and pipe fitting, but obviously it involves other types of tools, fittings, and controls. Unless this specialty is included in your apprentice program, you will have to get additional training after completing a plumbing or fitting apprenticeship. This training is available from industry associations and unions.

It is a growing field, so job opportunities are growing. Pay will compare favorably with plumbing and fitting.

Much of this trade is unionized. There also are a lot of nonunion shops. Much of the work, though, will be in large construction projects where frequently all work is unionized, so this choice will depend on where you want to work.

KITCHEN AND BATHROOM REMODELING

Kitchen and bath remodeling involves plumbing, heating, cooling, electricity, gas, and even carpentry and tile-setting.

But it is different in that the job can include tearing out the old, roughing in the new, and then finishing off so that the room looks like new. The job involves so much that kitchen and bath contractors very often move totally into this field and abandon any attempts to build business in the original trade.

It is an excellent field for plumbers, because plumbing usually is the main subcontract. This is especially true in bath remodeling. And this is why, historically, most kitchen and bath remodelers came from the plumbing field.

However, it is a separate and quite different business from plumbing. This is because the plumber usually is called to fix something that is broken. But the kitchen or bath remodeler must generate business by advertising and promotion and by doing such a good job that each customer brags about the remodeler.

It involves creative selling, because a new kitchen will cost as much as a new luxury car. A new bathroom will cost as much as a new compact car. Where the plumber would just be selling service, the remodeler will be selling new cabinets, appliances, lighting, floor and wall coverings, bath fixtures, counters, and even entertainment centers in either kitchen or bath.

That means each job will bring in many thousands of dollars. And you make much more money from a sale of several thousand dollars than you do from a $100 service call.

Nationwide there are about 10 million or more kitchens and bathrooms remodeled each year. This figure has been growing each year for the last 25 years, and chances are it will continue to grow.

So, because so many plumbers have entered bathroom and kitchen remodeling, it must be considered one of the most attractive related fields.

Some plumbers and pipe fitters work for big-job contractors as
members of large crews. (*SNIPS* magazine photo)

CHAPTER 8

SOME IMPORTANT ORGANIZATIONS

Anyone contemplating apprenticeship in plumbing or pipe fitting should be aware of the several organizations that have contributed to the growth and success of the plumbing industry and its related fields.

Some are contractor associations. Some are organized labor unions. Some are governmental or regulatory bodies. Here we'll look at some of those with which you will be most concerned.

UNION OR NONUNION?

Sooner or later you will have to consider union membership.

The union's apprentice program takes five years before you reach journeyman status, whereas other programs take four years or less. The meaningful requirements are not in years but in hours: 576 hours of classroom study and 8,000 hours of on-the-job training.

When you become a journeyman you will find three types of establishments in which you can apply for a job. They are known as open shops, union shops, and dual shops.

You do not have to join any union, but generally you

cannot get a job in a union shop unless you join. For instance, very large construction projects are often totally unionized. However, many states, especially in the South, now have "right-to-work" laws under which you must be hired if you qualify and if vacancies exist. Under right-to-work laws you can be either union or nonunion, and you might be working beside others who are or who are not union members.

The continuing battle between organized labor on the one hand and business and industry on the other is often bitter. Organized labor has lost a lot of membership in recent years, and the result often is a curious cross-pollination of specific unions and their local chapters. For example, in an area where one union is strong and another is weak, you might find the strong union organizing any group of workers regardless of what kind of work they do. Thus you might find a group of plumbers organizing as a local chapter of the Teamsters Union or the Teachers Union.

Unions generally are credited with gaining good pay scales, good working conditions, and many other benefits now enjoyed by all workers. Many say that all of these benefits now are facts of life and that the only effect of unionism now is to add to the costs of doing business.

But organized labor does remain a very potent force in Congress, in state legislatures, and in American business and industry. Every worker should investigate just what it can mean to him or her personally and then make a personal decision.

Following are the major unions and other organizations, listed alphabetically except for the first two, which are listed first because of their size and importance to a newcomer in the industry.

UNITED ASSOCIATION (UA)

VA Building
901 Massachusetts Avenue NW
Washington, DC 20001

This is the major national plumbers union. Its full name is the United Association of Journeymen and Apprentices of the Plumbing and Pipe Fitting Industry of the United States and Canada. It is commonly known as the UA—for obvious reasons.

The UA has hundreds of local chapters, many of them with their own buildings, training facilities, and instructors who conduct apprentice programs. This means, of course, that through the UA you would find your opportunities for training multiplied all over the country.

There's a catch to this, though. If you join the UA, you would have to start paying union dues right away.

NATIONAL ASSOCIATION OF PLUMBING-HEATING-COOLING CONTRACTORS (NAPHCC)

PO Box 6808
Falls Church, VA 22046
(703) 237-8100

This is the primary trade association for contractors in the plumbing, heating, and cooling industry, both large and small. Some do only plumbing, some do only cooling, some do all types of work in the pipe trades. As detailed in chapter 3, the association has complete training programs for both apprentices and contractor members, including video and home study courses and even a scholarship program. The home study program is especially valuable for apprentices in rural areas who can get on-the-job training with a local contractor but who might be far from any school and so cannot get the necessary classroom hours. With this course an apprentice in a rural area can complete the entire apprenticeship program and get a completion certificate.

Most contractor members of NAPHCC have open shops (60 percent). A little more than half that many, 36 percent, have union shops. Only 4 percent have dual shops. The total represents about 6,000 member firms who responded to a recent survey. The annual volume per firm was about $2 million, and the firms employed an average of 18 plumbers.

AMERICAN FIRE SPRINKLER ASSOCIATION (AFSA)

11325 Pegasus, Suite E-109
Dallas, TX 75238
(214) 349-5965

Fire sprinkler contractors that have open shops belong to this association. It is newer than the NFSA (listed later), but has an extremely comprehensive and aggressive training program for both apprentices and contractors.

ASSOCIATED BUILDERS AND CONTRACTORS, INC.

729 15th Street NW
Washington, DC 20005
(202) 637-8800

This nationwide group consists of about 20,000 general contractors, subcontractors, suppliers, and associates that believe in what they call the "merit shop" form of construction.

It was founded by six Maryland contractors in 1950. The objective was not to fight unionism but to promote merit shop construction wherein construction contracts would be awarded on the merits of the bidder rather than on whether the bidder was a union shop. According to ABC, in 1970 only 30 percent of the nation's construction was performed by

This manager of a plumbing wholesale company uses a computer to keep inventory records. (*Plumbing & Mechanical* magazine photo)

open shops. Now that figure has grown to more than 70 percent. ABC now has 78 chapters in 50 states, Canada, Guam, and the Virgin Islands.

To those contemplating a career in plumbing or fitting, the significant point is the ABC "Wheels of Learning" program, which offers complete courses for apprenticeship in plumbing and all related trades. But it also offers crossover courses in the related fields that might better enable the trainee to get a job that calls for specific skills outside plumbing. (See chapter 3 for details.)

ASSOCIATED SPECIALTY CONTRACTORS

7315 Wisconsin Avenue
Washington, DC 20014
(301) 657-3110

MECHANICAL CONTRACTORS ASSOCIATION OF AMERICA

5530 Wisconsin Avenue
Washington, DC 20088
(301) 897-0770

This association of contractors actually is in competition with the NAPHCC. It is generally thought of as an association of big-job contractors, although many of its members might engage only in plumbing and might be fairly small. Its educational programs and management materials are oriented more to contractors than to apprentices.

NATIONAL ASSOCIATION OF TRADE AND TECHNICAL SCHOOLS (NATTS)

Box 10429, Department CC
Rockville, MD 20850
(202) 333-1021

This is an association of technical and career schools nationwide, with quite a bit of career counseling. It gives the apprentice the opportunity for faster completion of classroom hours, but 8,000 hours on the job are still needed to complete an apprenticeship. NATTS publishes a catalog of courses offered in various schools in various trades.

NATIONAL FIRE SPRINKLER ASSOCIATION (NFSA)

PO Box 1000
Patterson, NY 12563
(914) 878-4200

This association of fire sprinkler contractors is all union. It has a complete apprenticeship training committee:

National Sprinkler Fitters Training Committee

7676 New Hampshire Avenue
Langley Park, MD 20783
(800) 638-0592

This committee is concerned with apprentices and their training, and it is integrally related to the National Fire Sprinkler Association. But it is in a different location and has its own director. This association and this committee consider an apprentice a union member right from the start. Accordingly, from the start of apprenticeship there are employer contributions in the apprentice's name into a welfare fund, a pension fund, and an apprentice educational fund.

Commercial plumbers and fitters usually work on regular routes, calling on customer firms and keeping their mechanical equipment in good condition. (*Plumbing & Mechanical* magazine photo)

CHAPTER 9

THE JOB CORPS AND OTHER AID

In 1988 there were two major federal funding programs to help young people get training and, in some cases, to help them get jobs.

The date is important because federal programs can change, particularly when Congress attempts to cope with budget deficits. Some programs get expanded. Some get dropped. Some undergo name changes. But in this chapter we'll tell you where and how to find out.

The two major programs are the Job Corps and the Job Training Partnership Act (JTPA).

JOB CORPS TRAINING CENTERS

The Job Corps was set up to help you get training when you find it difficult to get it elsewhere.

For example, it is difficult to get into a plumbing apprentice program if you don't have a high school diploma. The Job Corps is there to help in such cases, and it is there to help the underprivileged.

It has training centers all over the United States where you can be trained for an apprentice program in plumbing, pipe fitting, or dozens of other trades or for many other skills and crafts needed for productive work. In these training centers you can get a high school diploma or an equivalency certificate that can help qualify you for apprenticeship. A training center can help make up any deficiencies there might have been in your earlier school years.

Many thousands of young men and women join the Job Corps each year for such reasons. According to the U.S. Department of Labor, about 87 percent of these thousands are high school dropouts, and about one-third of them come from families who are living with the help of government programs. Almost 30 percent of them are women, and 70 percent represent minority groups.

And what they get from the Job Corps does help. According to Job Corps statistics, 20 percent of Job Corps trainees move into further education or training of some sort, and that includes apprenticeship programs in plumbing and pipe fitting.

About 15 percent join the armed forces, which in itself is a form of further education, because the armed forces have become very specialized and cannot afford untrained personnel. Many, many thousands of persons have been trained by the armed forces, including even such professionals as doctors.

The armed forces use scientific tests to discover the aptitudes of enlistees and then provide whatever training is necessary to develop the needed levels of skill in line with those aptitudes.

But the main point is that 80 percent of Job Corps trainees do find jobs after finishing the program.

Applicants for the Job Corps are 16 to 21 years of age. All candidates are screened, and their aptitudes are evaluated at the nearby Job Corps training center by a group that usually includes the director of the training center and an educational director.

It is at this point that as an applicant you should indicate an interest in a specific field, such as plumbing or pipe fitting. Then aptitude tests can be directed to calculate how well you probably can do in the field. If you do well in the tests, your training can be directed to the field you chose. If you don't do well, it does not necessarily close you out of it, especially if your interest is strong.

But from the tests you might find there is another field in which you could possibly do very well, and you might decide it would be to your liking. This all is done in consultation with you.

On acceptance in the corps you undergo a full year of training and instruction. Some of your work is in the classroom and some of it is in the field, actually performing the tasks of the chosen field. The classroom work will lead to a general equivalency diploma (GED), and the field work will develop job skills. In the field work real buildings are built, and you might be actually laying out and installing the piping for kitchens and bathrooms and the pipe or ductwork for heating and cooling systems.

This kind of work won't actually qualify you as a journeyman plumber. But it can qualify you for an apprenticeship, and the corps will help find you a job after this training is completed.

As a corps member, you get paid. The rate can change, but at this time it is $80 per month, and there are incentives that can swell that to $100 per month. You also get free room and board and "savings" in the form of a $100-per-month "readjustment allowance" that is given to you when you leave the center.

You can get more information on the Job Corps and a list of Job Corps Training Centers by writing the U.S. Department of Labor, Employment and Training Administration, Job Corps Information, Washington, DC 20213.

JOB TRAINING PARTNERSHIPS

This program uses federal funding, but it is administered and controlled by state governors.

It originated with the Job Training Partnership Act in 1983. The objective is to train unemployed workers so they can get jobs. In the states it is administered locally by private industry councils (PICs), and funds are supposed to be allotted to industry and labor groups to finance on-the-job training.

So the PICs are around, and they have money available. But they are not easy to find. The best idea is to call your state department of labor, listed in the white pages of the telephone book under the name of the state, and ask how to proceed.

If you can't get the information that way, the phone number in Washington, D.C., is (202) 535-0577. This gets you right into the office that has the names of all local PICs.

JTPA has a special training and employment program for veterans recently separated from military service, veterans of the Vietnam War, and veterans with service-connected disabilities. This program is Title IV-C, and you can get information on it from the assistant secretary for veterans employment and training in the U.S. Department of Labor.

There also is a special program for the National Congress of American Indians. For information on this program, contact the National Congress of American Indians, 804 D Street NE, Washington, DC 20002.

Governmental units from local towns on up to federal agencies are interested in employment and in training or re-training, so there might be new programs at any time. To find out about what is new, make it a point to check with local vocational-training school or community college directors and with any local chapters of unions of the trade in which you are interested.

VICA CLUBS

In many towns and cities, if you look in the Yellow Pages under "Schools—Industrial & Technical & Trade," you might find "VICA."

VICA stands for Vocational Industrial Clubs of America. Its members are students who are enrolled in vocational schools, community colleges, or high schools and who are interested in particular trades or occupations. There is a national VICA, and there are VICA associations for all the states plus Washington, D.C., Puerto Rico, and the Virgin Islands. Total membership is about 270,000 students in about 13,000 local clubs.

The idea here is to get together with other students in groups to promote your skills in the chosen field. VICA clubs are organized along trade or occupational lines, so if you are interested in plumbing or pipe fitting you will not be in a group whose members are interested in computers or some other different field. The groups hold local, state, and national competitions.

If you don't find a VICA in your community, you can get more information from the national office. Write VICA, Box 3000, Leesburg, VA 22075.

DOING IT YOURSELF

While help is available from various individuals and groups, the simple fact is that it is very easy to get into plumbing and pipe fitting on your own. There is plenty of room for more workers in the industry, and anyone who is able and willing to work can be hired.

You might, of course, have special problems. If, for example, you never completed high school, for whatever reason, you still should not give up hope. High school equivalency diplomas are available in just about any community. Check with the local school system. In a local school, this can be achieved in conjunction with on-the-job training. Then you can continue into apprenticeship training in a local vocational-technical school where your schooling is free.

And if you dropped out of high school years ago, remember, everyone learns a lot just by living and getting older. The high school diploma that was elusive years ago might come very easily now.

OFFICES OF THE BAT

The address for the national office of the Bureau of Apprenticeship and Training (BAT) is the following:

National Office
  Bureau of Apprenticeship and Training
  United States Department of Labor
  601 D Street NW
  Washington, DC 20213

However, many state and area offices of the bureau have been set up to help you start your apprentice training. These are parts of the federal system. In addition, some state and territorial governments have set up agencies of their own, and there might be some differences between the standards of these agencies and the national standards of the BAT.

The best advice is not to worry about it. The main thing is to get going. You can inquire about adapting to national standards, if necessary, after your first or second year, when you are well on your way.

In the following listing, the BAT's main state office is listed first with a telephone number. Other area offices are listed for your convenience, but without phone number.

Alabama
1931 Ninth Avenue South
South Twentieth Building
Birmingham, AL 35205
(205) 254-1308

U.S. Courthouse, Room 306
101 Homes Avenue
Huntsville, AL 35801

Room 418
951 Government State Building
Mobile, AL 36604

Alaska
Federal Building and Courthouse
701 C Street, Room E-512
Anchorage, AK 99513
(907) 271-5035

Arizona
2120 North Central
Suite G-110
Phoenix, AZ 85004
(602) 261-3401

Room 2-K, Federal Building
301 West Congress Street
Tucson, AZ 85701

Arkansas
Federal Building, Room 3014
700 West Capitol Street
Little Rock, AR 72201
(501) 378-5415

California
Room 344
211 Main Street
San Francisco, CA 94105
(415) 974-0556

Room 3235, Federal Building
300 North Los Angeles Street
Los Angeles, CA 90012

Room 6S-27, Federal Building
880 Front Street
San Diego, CA 92188

Colorado
United States Custom House
721 19th Street
Denver, CO 80202
(303) 837-4793

Connecticut
Federal Building, Room 367
135 High Street
Hartford, CT 06103
(203) 244-3886

Delaware
Federal Building, Box 36
844 King Street
Wilmington, DE 19801
(302) 573-6113

Florida
Hobbs Federal Building, Room 3080
227 North Bronough Street
Tallahassee, FL 32301
(904) 681-7161

Box 35082
400 West Bay Street
Jacksonville, FL 32202

Georgia
1371 Peachtree Street NE
Room 725
Atlanta, GA 30367
(404) 881-4403

Room 101
307 15th Street
Columbus, GA 31901

Room 236, Post Office Building
Box 8121
Savannah, GA 31402

Hawaii
300 Ala Moana Boulevard
Room 5113
Honolulu, HI 95850
(808) 546-7569

Idaho
1109 Main Street
Owyhee Plaza, #250
Boise, ID 83702
(208) 334-1013

Illinois
7222 West Cermak Road
Room 505
North Riverside, IL 60546
(312) 447-0382

Suite 36
3166 Des Plaines Avenue
Des Plaines, IL 60018

Suite 250
707 Berkshire Avenue
East Alton, IL 62024

Room 150, Federal Building
211 South Court Street
Rockford, IL 61101

Room 102, United States Post Office
600 East Monroe Street
Springfield, IL 62701

Indiana
Federal Building
46 East Ohio Street, Room 414
Indianapolis, IN 46204
(317) 269-7592

Suite 213, Riverside One
101 Court Street
Evansville, IN 47708

Room 342
1302 South Harrison Street
Fort Wayne, IN 45803

Room 210
610 Connecticut
Gary, IN 46401

Room 430, Sherland Building
105 East Jefferson Street
South Bend, IN 46601

Iowa
Federal Building, Room 367
210 Walnut Street
Des Moines, IA 50309
(515) 284-4690

Room 314-B, Federal Building
131 East Fourth Street
Davenport, IA 52801

Kansas
Federal Building, Room 367
444 Southeast Quincy Street
Topeka, KS 66683
(913) 295-2624

Suite 110, Page Court Building
200 West Douglas Street
Wichita, KS 67202

Kentucky
Federal Building, Room 554-C
600 Federal Place
Louisville, KY 40202
(502) 582-5223

Room 303
400 East Vine
Lexington, KY 40202

Louisiana
Hoover Building, Room 215-B
8312 Florida Boulevard
Baton Rouge, LA 70806
(504) 923-3431

Room 2502, Post Office Building
921 Moss Street
Lake Charles, LA 70601

925 South Street
600 F. Edward Herbert Building
New Orleans, LA 70130

Room 8A-09 Federal Building
500 Fannin Street
Shreveport, LA 71101

Maine
Federal Building, Room 101-B
68 Sewall Street
Augusta, ME 04330
(207) 622-8235

Maryland
Federal Building, Room 1028
31 Hopkins Plaza
Baltimore, MD 21201
(301) 962-2676

129 W. Main Street
Box 366
Salisbury, MD 21801

Massachusetts
JFK Federal Building, Room 1001
Government Center
Boston, MA 02203
(617) 223-6745

Room 211, Springfield Federal Building
1550 Main Street
Springfield, MA 01130

Room 500 Federal Building
United States Courthouse
Worcester, MA 01601

Michigan
Room 308, Corr Building
300 East Michigan Avenue
Lansing, MI 48933
(517) 377-1746

Room 2-1-60
Battle Creek Federal Center
74 North Washington Avenue
Battle Creek, MI 49107

Room 657, Federal Building
231 West Lafayette Avenue
Detroit, MI 48226

Room 186, Federal Building
110 Michigan NW
Grand Rapids, MI 49502

Suite 410, City Hall
220 West Washington Street
Marquette, MI 49855

North Warren at East Genesee
Box 1017
Saginaw, MI 48606

Minnesota
Room 134, Federal Building
316 Robert Street
Saint Paul, MN 55101
(612) 725-7951

Room 235, Federal Building
515 West First Street
Duluth, MN 55802

Mississippi
Federal Building, #1003
100 West Capitol Street
Jackson, MS 39269
(601) 960-4346

Missouri
210 North Tucker, Room 547
Saint Louis, MO 63101
(314) 452-4522

Room 2111 Federal Office Building
911 Walnut Street
Kansas City, MO 64106

Montana
Federal Office Building, #394
301 South Park Avenue
Helena, MT 59626
(406) 449-5261

Room 1414 Federal Building
316 North 26th Street
Billings, MT 59101

Nebraska
106 South 15th Street
Room 700
Omaha, NE 68102
(402) 221-3281

Nevada
Post Office Building, #316
301 East Stewart Avenue
Las Vegas, NV 89101
(702) 385-6396

New Hampshire
Federal Building, #311
55 Pleasant Street
Concord, NH 03301
(603) 834-4736

New Jersey
402 East State Street, #410
Trenton, NJ 08607
(609) 989-2209

Room 838 New Federal Building
970 Broad Street
Newark, NJ 07102

New Mexico
Western Bank Building, #1116
505 Marquette NW
Albuquerque, NM 87102
(505) 766-2398

New York
512 United States Post Office and Courthouse
Albany, NY 12207
(518) 472-4800

Room 303, 15 Henry Street
Box 308
Binghamton, NY 13902

Room 220, Federal Building
111 West Huron Street
Buffalo, NY 14202

Room 3731
1515 Broadway
New York, NY 10036

Room 607, Federal Building
100 State Street
Rochester, NY 14614

Room 1241, Federal Building
100 South Clinton Street
Syracuse, NY 13202

North Carolina
Federal Building, #376
310 New Bern Avenue
Raleigh, NC 27601
(919) 755-4466

Room 410, Merchandise Mart Office Building
600 Briar Creek Road
Charlotte, NC 28202

Ohio
200 North High Street
#605
Columbus, OH 43215
(614) 469-7375

Room 208 Federal Building
201 Cleveland Avenue SW
Canton, OH 44702

Room 2112, Federal Office Building
550 Main Street
Cincinnati, OH 45202

Room 720, Plaza 9 Building
55 Erieview Plaza
Cleveland, OH 44114

Room 613, Federal Building
200 West Second Street
Dayton, OH 45404

Room 706, Federal Office Building
234 Summit Street
Toledo, OH 43604

Room 812, City Centre One
100 Federal Plaza East
Youngstown, OH 44513

Oklahoma
50 Penn Place, #1440
Oklahoma City, OK 73118
(405) 231-4818

Suite C, Center Mall Professional Building
222 South Houston Avenue
Tulsa, OK 74127

Oregon
840 Federal Building
1220 Southwest Third Avenue
Portland, OR 97204
(503) 221-3157

Room 231 Federal Building
211 East Seventh
Eugene, OR 97401

Pennsylvania
Federal Building, #773
228 Walnut Street
Harrisburg, PA 17108
(717) 782-3496

Room 106 Federal Building
Sixth and State Streets
Erie, PA 16507

Room 13240, Gateway Building
3535 Market Street
Philadelphia, PA 19104

Room 1436, Federal Building
1000 Liberty Avenue
Pittsburgh, PA 15222

Room 2115, East Shore Office Building
45 South Front Street
Reading, PA 19603

Room 2028
20 North Pennsylvania Avenue
Wilkes-Barre, PA 18701

Rhode Island
100 Hartford Avenue
Providence, RI 02909
(401) 838-4328

South Carolina
Strom Thurmond Federal Building, #838
1835 Assembly Street
Columbia, SC 29201
(803) 765-5547

Room 231, Federal Building
344 Meeting Street
Charleston, SC 29403

South Dakota
Federal Building, #104
400 South Phillips Avenue
Sioux Falls, SD 57102
(605) 336-2980

Tennessee
1720 West End Avenue
Suite 406
Nashville, TN 37203
(615) 251-5408

6300 Building, Suite 7003
Eastgate Center
Chattanooga, TN 37411

Room 232
301 Cumberland Avenue
Knoxville, TN 37902

Room 209, Federal Office Building
167 North Main Street
Memphis, TN 38103

Texas
VA Building, #2101
2320 LaBranch Street
Houston, TX 77004
(713) 750-1696

Room 325, Federal Building
300 Willow Street
Beaumont, TX 77701

Suite 306
1403 Slocum
Dallas, TX 75207

Room 9A08, Federal Building
819 Taylor Street
Fort Worth, TX 76102

Room 416 Federal Building
1205 Texas Avenue
Lubbock, TX 79401

Room B, 414 Federal Building
727 East Durango
San Antonio, TX 78206

Utah
Post Office Building, #314
350 South Main Street
Salt Lake City, UT 84101
(801) 524-5700

Vermont
Burlington Square
96 College Street, #103
Burlington, VT 05401
(802) 951-6278

Virginia
400 North Eighth Street
Room 10-020
Richmond, VA 23240
(804) 771-2488

Room 426, Federal Building
200 Granby Mall
Norfolk, VA 23510

Washington
1009 Federal Office Building
909 First Avenue
Seattle, WA 98174
(206) 442-4756

545 United States Courthouse
West 920 Riverside
Spokane, WA 99201

Suite 50, Old City Hall
625 Commerce Street
Tacoma, WA 98402

West Virginia
550 Eagan Street
Room 305
Charleston, WV 25301
(304) 347-5141

Room 2701, Federal Building
425 Juliani Street
Parkersburg, WV 26101

Wisconsin
Federal Center, #303
212 East Washington Avenue
Madison, WI 53703
(608) 264-5377

Prairie Professional Park
6921 Mariner Drive
Racine, WI 53406

Room 128, Wood County Courthouse
400 Market Street
Wisconsin Rapids, WI 54494

Wyoming
O'Mahoney Federal Center, #8017
2120 Capitol Avenue
Cheyenne, WY 82001
(307) 772-2448

There are state and territorial apprenticeship agencies in 29 states and in Puerto Rico and the Virgin Islands. The federal BAT encourages states to set up these supplemental agencies, but since they are not under federal authority there is no assurance of compliance with BAT rules.

Arizona
Apprenticeship Services
Department of Economic Security
207 East McDowell Road
Phoenix, AZ 85004

California
Division of Apprenticeship Standards
Department of Industrial Relations
455 Golden Gate Avenue, Room 3230
San Francisco, CA 94102

Colorado
Apprenticeship Council
Division of Labor
323 Centennial Building
Denver, CO 80203

Connecticut
Apprentice Training Division
Department of Labor
200 Folly Brook Boulevard
Wethersfield, CT 06109

Delaware
Apprenticeship Officer
Delaware State Department of Labor
State Office Building, Sixth Floor
820 North French Street
Wilmington, DE 19801

District of Columbia
DC Apprenticeship Council
500 C Street NW, Room 241
Washington, DC 20001

Florida
Bureau of Apprenticeship
Division of Labor
Department of Labor and Employment Security
1321 Executive Center Drive East
Tallahassee, FL 32301

Hawaii
Apprenticeship Division
Department of Labor and Industrial Relations
825 Mililani Street
Honolulu, HI 96813

Kansas
Apprenticeship Section
Division of Labor-Management Relations and
    Employment Standards
Department of Human Resources
512 West Sixth Street
Topeka, KS 66603

Kentucky
Apprenticeship and Training
Kentucky State Apprenticeship Council
620 South Third Street, Sixth Floor
Louisville, KY 40202

Louisiana
Division of Apprenticeship
Department of Labor
PO Box 44094
Baton Rouge, LA 70804

Maine
  State Apprenticeship and Training Council
  Department of Manpower Affairs
  Bureau of Labor
  State Office Building
  Augusta, ME 04333

Maryland
  Apprenticeship and Training
  Maryland Apprenticeship and Training Council
  Division of Labor and Industry
  Baltimore, MD 21237

Massachusetts
  Division of Apprentice Training
  Department of Labor and Industries
  Leverett Saltonstall Building
  100 Cambridge Street
  Boston, MA 02202

Minnesota
  Division of Voluntary Apprenticeship
  Department of Labor and Industry
  Space Center Building, Fifth Floor
  444 Lafayette Road
  Saint Paul, MN 55101

Montana
  Apprenticeship Bureau
  Division of Labor Standards
  Department of Labor and Industry
  Capitol Station
  Helena, MT 50620

Nevada
  Department of Labor
  Capitol Complex, Room 601
  505 East King Street
  Carson City, NV 89710

New Hampshire
  Commission of Labor
  Department of Labor
  19 Pillsbury Street
  Concord, NH 03301

New Mexico
New Mexico State Apprenticeship Council
Labor and Industrial Commission
2340 Menaul NE, Suite 212
Albuquerque, NM 87107

New York
Apprentice Training
Department of Labor
Campus Building 12, Room 428
Albany, NY 12240

North Carolina
Apprenticeship Division
North Carolina Department of Labor
4 West Edenton Street
Raleigh, NC 27601

Ohio
State Apprenticeship Council
Department of Industrial Relations
2323 West Fifth Avenue, Room 2250
Columbus, OH 43215

Oregon
Apprenticeship and Training Division
State Office Building, Room 466
1400 Southwest Fifth Avenue
Portland, OR 97201

Pennsylvania
Apprenticeship and Training Council
Department of Labor and Industry
Labor and Industry Building, Room 1618
Seventh and Forster Streets
Harrisburg, PA 17120

Puerto Rico
Apprenticeship Division
Department of Labor
Right to Employment Division
GPO Box 4452
San Juan, PR 00936

Rhode Island
Apprenticeship Council
Department of Labor
220 Elmwood Avenue
Providence, RI 02907

Vermont
Vermont Apprenticeship Council
Department of Labor and Industry
120 State Street
Montpelier, VT 05602

Virgin Islands
Division of Apprenticeship and Training
Department of Labor
Christiansted
Saint Croix, VI 00820

Virginia
Division of Apprenticeship Training
Virginia Department of Labor and Industry
PO Box 12064
Richmond, VA 23241

Washington
Apprenticeship and Training Division
Department of Labor and Industries
605 East 11th Avenue
Olympia, WA 98504

Wisconsin
Division of Apprenticeship and Training
Department of Industry, Labor, and Human Relations
PO Box 7946
Madison, WI 53707

WHERE TO GET MORE INFORMATION

ORGANIZATIONS

American Fire Sprinkler Association
11325 Pegasus, Suite E-109
Dallas, TX 75238

Employment and Training Administration
United States Department of Labor
200 Constitution Avenue NW
Washington, DC 20210

Mechanical Contractors Association of America
5530 Wisconsin Avenue
Washington, DC 20088

National Association of Plumbing-Heating-Cooling Contractors
PO Box 6808
Falls Church, VA 22046

National Association of Trade and Technical Schools
2251 Wisconsin Avenue NW
Washington, DC 20007

National Kitchen and Bath Association
124 Main Street
Hackettstown, NJ 07840

National Sprinkler Fitters Training Committee
7676 New Hampshire Avenue
Langley Park, MD 20783

PUBLICATIONS

Apprenticeship: Past and Present. A history and description of apprenticeship in the United States. Free from the U.S. Department of Labor, Employment and Training Administration, 601 D Street NW, Washington, DC 20213.

Building America. Catalog of services, programs, and training courses available from the Associated Builders and Contractors, Inc. Free.

Construction and Extractive Occupations, Bulletin 2250-18. Booklet describing plumbing, piping, sheet metal, and related occupations. Published by the U.S. Department of Labor, Bureau of Labor Statistics.

National Apprenticeship Program. Pamphlet. Available from the U.S. Department of Labor, Bureau of Apprenticeship and Training, 601 D Street NW, Washington, DC 20213.

Pipe Fitter's and Pipe Welder's Handbook. Published by Macmillan Company, Riverside, NJ 08075. $5.75.

Pipe Fitters and Steam Fitters, Brief 285. Four-page description of these occupations. Published by Chronicle Guidance Publications, Box 1190, Moravia, NY 13118. $3.

Piping Pointers. A free book published by Crane Canada, Box 2700, St. Laurent, Montreal 9, Quebec.

Plumbers, Brief 43. Four-page description of the field published by Chronicle Guidance Publications, Box 1190, Moravia, NY 13118. $3.

Sheet Metal Workers, Brief 222. Pamphlet published by Chronicle Guidance Publications, Box 1190, Moravia, NY 13118. $3.

Woman's Guide to Apprenticeship. A free pamphlet available from the U.S. Department of Labor, Women's Bureau, Washington, DC 20210.

Your Future in the Plumbing-Heating-Cooling Industry. Pamphlet. Available from National Association of Plumbing-Heating-Cooling Contractors, P.O. Box 6808, Falls Church, VA 22046.

GLOSSARY OF COMMON PLUMBING TERMS

Air chamber.   Mechanical device that reduces water pressure in a water line. Air chambers may be installed at a fixture or at the service main, and they usually are 12- to 18-inch capped extensions of the water supply lines. They are usually in the wall near the fixture, providing air cushions for high pressure. They belong in all pipes bringing cold and hot water to the bathroom group.

Air hammer.   Pounding in a water supply line that can occur when a faucet is shut off rapidly. It usually happens when water pressure is more than 60 pounds per square inch and is caused by air in the pipe. It can be eliminated by an air chamber.

Apprentice.   A student of any trade who has enrolled in a formal apprenticeship training program that can lead to a completion certificate, which takes 8,000 hours of on-the-job training plus 576 hours of classroom work over four or five years in plumbing or pipe fitting.

Back-siphonage.   The force caused by negative pressure in supply pipe or higher pressure in waste line that causes contaminated water to flow back into potable water supply.

Bathroom group.   Standard grouping of fixtures in a bathroom, consisting of lavatory, toilet, bathtub, or shower stall.

Blueprint.   A copy of the detailed architectural plans of a structure or of a kitchen or bathroom. It is usually blue in color because of the copying process involved.

Boiler.   Heating appliance that might be gas-, oil-, or electric-fueled, for heating water in a hydronic heating system.

Brass.   Collective term for faucets in kitchen and bathroom. The term has nothing to do with the material from which faucets are made.

Code.   Regulations and rules written by federal, local, or state governments or by another code-writing body, to control materials and installation procedures to protect the health, safety, and welfare of the public.

Compression fitting.   A fitting for joining lengths of copper tubing without solder or torch. Compression fittings usually are made of brass and are threaded. When tightened they compress the soft copper to make a seal. (See sweat fitting.)

Continuous waste.   A drain from two or more fixtures connected to a single trap.

Duct and Ductwork.   Piping fabricated of sheet metal to distribute hot or cold air in heating and air-conditioning systems.

DWV.   Common term for drain-waste-vent, used to describe the piping system that carries waste away.

Fittings.   Devices used to join pipes. Common fittings are nipples, couplings, unions, elbows, tees, wyes.

Fixture.   Plumbing appliance commonly used in the bathroom, including lavatory, toilet, tub.

Floor plan.   Drawing of the layout of a kitchen or bathroom showing all fixtures, cabinets, appliances, doors, windows, lighting, and other equipment, to scale, as seen from above.

Grade or pitch.   The angle of slope for a waste line that permits gravity flow.

Furnace.   Heating appliance for central heating systems. It heats air that then is distributed throughout the house by a blower at the furnace.

Hydronic heat.   Heating system in a house or other building in which heat is distributed by hot water or steam rather than warmed air. The hot water or steam might go to radiators in the rooms or through copper tubing in floors, walls, or ceilings to warm by radiant heat.

Journeyman. Trades person who has completed apprenticeship and is sufficiently skilled to do routine jobs unsupervised.

Master plumber. A skilled journeyman plumber who is licensed by a local licensing authority. Licensing usually concerns complete knowledge of the local plumbing code.

Pipe. Any tube with the purpose of carrying fluids, solids, or gases. It might be of iron, steel, plastic, concrete, or other material.

Pipe fitter. A plumber who has completed more training and works more in laying out and installing piping systems. This work might be in petroleum or chemical plants, or in certain large commercial jobs where much piping is involved.

Plastic pipe. Supply or DWV pipe made of any of various plastics. It is lighter in weight and much easier to handle than metal pipe, but it is not fully approved by codes in some areas.

Potable water supply. Water free enough of impurities to be safe to drink.

psi. Common abbreviation for pounds per square inch of pressure.

Revent. A pipe running upward from a fixture waste pipe and connecting it to the stack above all other waste connections. It is an alternative to putting in a new stack through the roof when the new fixture is too far from the stack for venting otherwise.

Riser. Vertical water pipe.

Rough-in. The plumbing for water supply and waste that must be installed behind the walls before the walls are finished so that bath, kitchen, or laundry fixtures can be installed after walls and floors are completed.

Service main. Piping that brings water supply from outside source to a building.

Soil pipe. Pipe that carries waste from toilet bowls to the soil stack.

Space heater. Heating device for small area.

Sprinkler fitter. A plumber who has had extra training and who installs and maintains automatic fire sprinkler systems.

Stack. Vertical pipe, usually 3 to 4 inches in diameter, that carries waste to the building drain, also extending upward to above the roof of the building for escape of gases.

Steam fitter. A pipe fitter who works with hot water and steam piping systems.

Supply line.   Pipe bringing the water supply under pressure from a main into a building, or distributing it through the building.

Sweat fitting.   Joining copper tubing and its fittings by soldering, a practice in which the copper, heated with a torch, "slurps" the solder into and throughout the joint by capillary action, making a total seal. (See compression fitting.)

T-P Valve.   Temperature and pressure relief valve on water heater as a safety measure to relieve excessive pressure in tank.

Trap.   Bend in a waste pipe used to create a water seal against backflow of sewer gases into a building. Common are U-traps, P-traps, and drum traps used in floor under toilet.

Trap seal.   The amount of water in a trap between the top and bottom of the bends, necessary to prevent entry of sewer gases into room.

Tubing.   Pipe made of copper.

Unit heater.   Heating appliance that provides large volume of hot air in a specific place, such as at a point just inside an outside doorway.

Vent pipe.   Pipe allowing air into the DWV system. It equalizes air pressure in the lines to prevent siphonage of water from traps.

Waste line.   Pipe carrying waste of any kind away from a plumbing fixture or building. Flow is by gravity rather than pressure, so it must slope downward in the direction of flow. When the waste line carries human waste, it is called a soil pipe.

Whirlpool tub.   Bathtub with motor and several inlets in sides through which air is jetted in below water surface. Its purpose is to relax the muscles of the bather.

A SELF-TEST FOR A CAREER
IN PLUMBING AND/OR
PIPE FITTING

Throughout this book, plumbing and pipe fitting careers have been described so that you can decide whether you are able to work in this field, might like it, and are likely to find it satisfying and rewarding.

In short, would this career suit you? Would you suit it? These are very important questions, because you don't want to put in years (or even months, for that matter) doing the preparatory work and getting into apprenticeship and ultimately into plumbing and pipe fitting only to find out that you do not like it at all or that you simply aren't suited to it.

To give you further help in your evaluation, here are some final questions and answers, a self-test based on knowledge gained from working in the field and the people who work in it.

At the right of the page, rate yourself on a scale of 1 to 5, with 1 being "very poor" or "not at all," 3 being "about average," and 5 being "excellent" or "very much." At the end, instructions will be given on how to rate yourself and what your score might mean for your future career in plumbing and pipe fitting.

129

1. Are you mechanically inclined?   1 2 3 4 5

Wait. Don't answer too quickly, not until you are sure what that means. Do you like to work on old cars, or do you like to try to fix a faucet that leaks or a drain that is stopped up? If a lawn mower needs sharpening, can you do it yourself? Or if the lock on the door doesn't work, or for any reason you find a hole in the wall, can you figure out something to do to correct it? Do you enjoy doing this kind of thing?

Notice the question is not about how well you do this kind of work. Rather, the question is about how much you like to do it.

Some people might not be able to do mechanical jobs well simply because they have never had an opportunity to see how things work or because they have never had the right tools or the time. If that includes you, think carefully and thoroughly about whether you would really like to do these kinds of jobs. Think of a few imaginary problems and what you might do to fix them. This might help you decide whether you have this aptitude, regardless of your background.

2. Are you eager to learn new things?   1 2 3 4 5

This is another question that you might find difficult to answer, because, after all, you are considering a new period in your life.

There are many young people who dislike school. This does not necessarily mean they are not eager to learn. Many times young people cannot see any real value in school, regardless of what their teachers or parents tell them. Whether this is a good or bad attitude is beside the point. The point is that it is a fact, and very few people want to learn anything that they see as having no value.

But now you are thinking about something entirely new, a

way to make a good living for the rest of your life. It will involve school, lots of it, but this will be schoolwork with a purpose, and you can see what that purpose is. You can see that it is worth the time and effort. So consider this before you answer. You have read this book. If you found plumbing and pipe fitting interesting, that might mean you would enjoy learning more about it.

3. Do you like physical activity?   1 2 3 4 5

This is important, because there is a lot of physical activity in any of the construction trades and in any of the plumbing and fitting subspecialties. It is hard and demanding work, and there is no place in it for people who do not like to exert themselves physically.

So if you tend to prefer work requiring less physical activity, you should give a career in plumbing and fitting some second thoughts. Plumbers and fitters often must lift heavy pipes and stand for long periods. Some may have to work evening or weekend shifts. Often they work in cramped, dirty places, and they may have to work outdoors in bad weather. There is even some risk of injury, as plumbers and fitters often have to work with ladders, sharp tools, and hot pipes.

So you should think about this question. If you are inclined to work indoors behind a desk, you probably should choose a career other than plumbing and pipe fitting.

4. Are you reasonably strong?   1 2 3 4 5

The plumbing trade requires a lot of heavy lifting, pulling, and pushing, and plumbers often have to work on their feet for long periods of time. Often they have to do a lot of heavy work on their knees or on their backs. They work with large, heavy, awkward-sized objects.

To do this work you don't have to have superhuman strength.

However, there are a lot of persons who are not strong enough for this work and who, for whatever reason, might not be capable of building up their physical strength; they probably should not pursue careers in plumbing or pipe fitting.

In considering this question, note also that you would seldom work alone. You nearly always would have a helper or be a helper. You simply have to be strong enough to hold up your end.

5. Are you nimble, spry, quick?   1   2   3   4   5

In addition to lifting, pulling, and pushing, plumbers also have to climb ladders, carry things up ladders, and work in high places both indoors and out. Pipe and steam fitters always risk burns from hot pipes or steam, and both plumbers and fitters sometimes work with torches.

This means any clumsiness can be dangerous both to the plumber and to others he or she is working with. Many people seem to have more minor accidents than others and are considered accident prone. They always seem to be cutting or burning or bruising themselves. This is a dangerous trait in plumbing and pipe fitting.

In answering this question, however, it is important to your career to realize that such accidents often are caused not by clumsiness but by undetected problems with seeing or hearing. Many persons have been "cured" of clumsiness by a new set of glasses or a hearing aid.

6. Do you figure things out yourself?   1   2   3   4   5

In all walks of life, even when you are very young, there are jobs to be done that you just don't understand immediately. In school it might be a test question that is worded in a complicated way, or it might be that a lawn mower is cutting the grass too high. Life is full of these types of challenges.

You have two choices in these matters. You can take the easier way and ask somebody immediately, or you can look at the problem again and try to figure it out for yourself before you seek help. In most cases, there is an answer if you just study the problem.

The important thing is not so much that you solve the problem but that you try to solve it without asking directions first, that you look at it in as many ways as you can. If you are able to solve problems, that is a good quality. But knowing when to seek help is a good sign for success in any career.

7. Do you often have to do things over? 1   2   3   4   5

Doing things wrong and having to do them over might be a problem you need to work on. It might be that you try to do things beyond your experience or knowledge, and that is good as long as you are doing things that aren't dangerous. But more often having to do things over is a sign of carelessness, or that you are not paying attention to what you are doing.

In plumbing and pipe fitting, this carelessness can be very hazardous to yourself and to others who are working with you. You might be the person responsible for erecting a scaffolding on which several persons will work, and a careless mistake could send others plunging to their deaths. In pipe or steam fitting you might be working with steam lines or gas lines, and careless mistakes can mean death or serious injury.

So this question relates to doing things carefully and thoroughly. If you are careful and thorough in whatever you do, score yourself high. If you make mistakes simply because you are enterprising or careless, you may want to work on becoming more careful before you pursue a career in plumbing and pipe fitting.

8. Do you wonder how things work?   1 2 3 4 5

Did you ever wonder how a clock works and even take it apart to watch it and test your theory? Did you ever fiddle with a 10-speed bike and try to figure out just what happens to make it change gears?

If your answer is "yes," it means that you are curious. Curiosity of this kind is valuable in any career and especially in construction trades such as plumbing and pipe fitting. It is the kind of trait that helps you solve problems on the job, making you more valuable to your employer, meaning promotions and increased pay for you.

It might be that you never figured out the secret of the ticking clock or the changing gears. That is not as important as the fact that you were curious enough to wonder about and to tinker with them. Experience, plus simply growing older, will help you solve problems, but first you have to be curious about them.

9. Do you get impatient easily?   1 2 3 4 5

Impatience often is confused with carelessness, but they are not the same. You might normally be careful and thorough, but if you tend to be impatient, you might find that you often start to speed up even when you don't want to. You will look for shortcuts, and that can lead to mistakes that endanger yourself and others.

This does not mean shortcuts are bad. Actually, they are very good, but only when they have been thought out carefully and thoroughly. True shortcuts can be better ways to get things done, and if you are good at finding them, you will be rewarded.

An act of impatience happens when you are actually on the job. It is something you do impulsively, usually when you become exasperated because whatever you are doing is progressing too slowly to suit you. Shortcuts, on the other hand, are procedures or methods that you figure out away from the job, perhaps even after you have gone home or before you start to work the next day.

10. Are job benefits important to you?   1   2   3   4   5

This question may seem silly, but the fact is that some people really don't care about their pay scale or other job benefits. They often do not put in the extra effort required for advancement in their jobs. So they float from job to job.

Job benefits in plumbing and pipe fitting or any of their subspecialties include good pay, steady work, opportunity for advancement, continual challenge on the job, variety, and always the chance to go out on your own and start your own business.

These are very worthwhile benefits, but each person must earn these benefits for him or herself by being productive and constructive. Earning these benefits requires hard work and hard study for four or five years, but the rewards last a lifetime.

So rate yourself on the kind of plumber or fitter you want to be. This trade is available to you, and it puts your future in your own hands.

Now add up your score. There are 10 questions with a maximum score of 5 each, so the best score you can get is 50. Here is how to rate yourself:

40 to 50—This makes you an excellent prospect for this kind of work. It doesn't guarantee success, of course. No preexamination can. You will still have to do the work and the hard studying. But if you have answered as honestly as you can, you are a type that any contractor will like.

35 to 40—If your score is in this range, it means you are a

very good prospect, and you can be reasonably sure of success.

25 to 35—A score in this range means you will really have to apply yourself to be good.

Under 20—You have a tough row to hoe, and it will take a lot of determination and hard work to succeed. If you are capable of that, then you have a good chance. Look back over the questions again and make sure you have given yourself a fair break on all of them. Make sure you really understood them.

Remember, even the most scientific test can be wrong. The only real test is to go ahead and try. To help you get a little better feel for this whole plumbing and pipe fitting field, see how comfortable you are as you read the next few pages.

How well do you like mathematics? If you read the various lesson plans in chapter 4, you find that mathematics is one of the important aspects of this trade. What does it have to do with plumbing and pipe fitting?

Some of the manuals of the *Wheels of Learning* program of the Associated Building Contractors answer this question. This program is described in chapter 3. All of the following material is from the *Pipefitting II Student Manual,* and it will give you an idea of the importance of basic mathematics to construction and how you will use it every day.

For example, if you are installing a water supply line and have to change direction to avoid a gas or sewer line, you will have to use an ancient mathematical rule called the Pythagorean theorem to establish the angles of offset and distances. If you have to install a guy wire to hold a tall pole, you will again use the Pythagorean theorem to determine the amount of wire you will need. Here is how.

First, there are several kinds of triangles, but all triangles have three sides and three angles. They are classified either by the characteristics of their sides or their angles, but we'll skip that and talk about one type, the right triangle, which has one angle that is 90°. As you may know, a 90° angle is called a right angle. It is a square corner.

You also should know that the angles of every triangle, without exception, add up to 180°. This is because every triangle is exactly one-half of a 4-sided figure called a parallelogram. By definition, a parallelogram always has 360°; therefore, half of it must contain 180°. This means that if you know two of the angles, you can always figure out the third angle simply by adding the two that you know and subtracting that sum from 180. If it is a right triangle, you always know that the square corner is 90°, so the other two angles must total 90°. The side of the triangle that is opposite the right angle is called the hypotenuse (see figure 1). You also should know that if you straighten out the two sides of any angle to make a straight line, that angle is 180°.

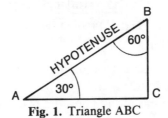

Fig. 1. Triangle ABC

What about that theorem? Many centuries ago in Greece, a mathematician named Pythagoras discovered special relationships among the sides and angles of a right triangle. His theorem, or rule, is that in a right triangle, the square of the hypotenuse is equal to the sum of the squares of the other two sides.

You can prove it yourself. The square of a number is that number multiplied by itself. Now look at figure 2. This is a right triangle with one side 3′ long, one side 4′ long, and the hypotenuse (the side opposite the right angle) measuring 5'. Here the squares are filled in to show the proof: The square of side AB (3 × 3) is 9, the square of side BC (4 × 4) is 16, and the square of the hypotenuse AC (5 × 5) is 25. Twenty-five is equal to the square of AB (9) plus the square of BC (16).

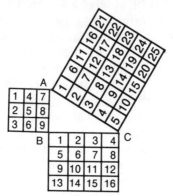

Fig. 2. Proof of the
Pythagorean theorem.

Fine. But what use is this to pipe fitters? What applications can it have in this field?

Plenty. Pythagoras didn't know it 2,600 years ago, but he was helping to figure piping offsets he never saw for a future he could never imagine.

We have already said that if you know any two angles, you can easily figure out the third. But you also can figure out an unknown side of a right triangle if you know two sides. In the mathematics field, mathematicians can talk about sides and the hypotenuse, but to a pipe fitter it boils down to run, offset, and travel.

Look at a right triangle as it would be formed in a piping offset in figure 3. Notice that all dimensions in piping offsets are measured from the center of the pipe. The various parts of the right triangle in this offset are defined as follows:

1. *Run* is the straight-line direction in which the pipe is advancing.
2. *Offset* is the center-to-center difference between the locations of the lines of pipe as they advance.
3. *Travel* is the distance required to span the offset and to connect the separate lines of pipe.

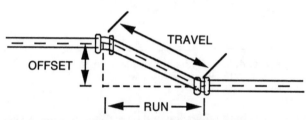

Fig. 3. Piping offset, measured with Pythagorean theorem.

You might have to know the length of pipe you will need for the travel. Or you might have to know the fittings you will need for the proper angles to continue the run from one level to the other. In any case, you can measure the distance of offset, then the distance to continue the run, and calculate the length of pipe for the travel. Or, having the right angle at the bottom of the offset, you can calculate the angles of the two fittings for the travel pipe. Remember, the offset might be a foot in a residential plumbing job or a mile in an oil refinery. You can't always just pull out a tape rule and measure it.

For example, look at figure 4. Here you have a 30′ pole that you have to secure with a guy wire. There is a wall or a place in the ground 40′ away where you can secure the guy wire. How long a wire do you need?

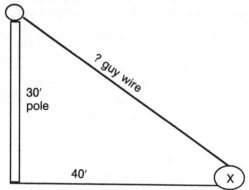

Fig. 4. Use Pythagorean theorem to calculate length of guy wire.

Many times in the field you will have to make a rough sketch of a problem that confronts you. The sketch will help you think the problem through and enable you to visualize the situation more clearly. Sketching is a very useful skill to develop.

The first step in sketching a problem is to draw on paper the general shape of the problem and label each known dimension. This clearly isolates the unknown dimension, so it can help you determine your approach to the problem.

Your sketch of the guy wire problem will look like figure 3. You can see clearly that is a right triangle problem. The length of the guy wire is the hypotenuse of the right triangle formed by the pole and the point where you will anchor the wire. So you know that the guy wire squared (the hypotenuse) will equal the sums of the height of the pole squared plus the distance to the anchor point squared. In other words, 30 squared (30×30) is 900, and that plus 40 squared (1600), or 2500, is equal to the length of the guy wire squared. If the guy wire squared is 2500, you only have to figure the square root of 2500 (what number times itself equals 2500), and that number is 50. So you need 50 feet of guy wire.

We are not trying to teach a mathematics lesson here, but perhaps the term *square root* should be explained. Generally, as a pipe fitter or plumber, you will have a pocket calculator as part of your equipment. It will have a square root function button, so you simply enter the number and push that button. You will know all of this, of course, by the time you finish apprenticeship training.

The Pythagorean theorem is just as useful and is used as much in house construction as out in the field. For example, every kitchen designer and installer and every manufacturer of custom kitchen and bathroom countertops use it every day.

They use it in a simplified, shorthand version. They may never have heard of a Greek named Pythagoras, but they all are very familiar with what they might call the 3-4-5 rule.

Their problem is this. In any house, new or old, no corner is truly square. Every corner will be square in the architect's drawing, but actually there is wood warpage and sheetrock and sheetrock taping. So a wall that is supposed to be 60″ long might be 60 1/2″ long at the wall between the corners but as much as an inch longer or shorter at the front of the countertop, which is 24″ from the wall. If you have ordered a 5′ countertop to fit here, you have a serious problem with both the counter manufacturer and the homeowner.

The rule of 3-4-5 is this: After measuring the wall as it really is, you measure and mark a line 3′ out from the corner on one wall and then measure and mark a line 4′ out from the corner on the other wall. Then you measure precisely between your marks. This, as you should recognize by now, is the hypotenuse of a right triangle. If you have a perfectly square corner, the distance between marks will be exactly 5′. If it is less, that is how much the wall comes inward. If it is more, that is how much the wall goes outward. You then order the countertop accordingly, so it will fit the corners properly.

As you can see, with this simplified rule there is no squaring of the sides or of the hypotenuse of the right triangle. But it works every day in every city in this country, and it is a necessary application of the old theorem by Pythagoras.

Now, how about exercising your mind a little on all of this. Look at figure 5 and calculate the number of degrees in the angles A, B, and C. No, we're not giving the answers. But notice: What you see is a parallelogram, which, in this case, happens to be a rectangle. That means each corner is a right angle. And you know already that there are 180° in every triangle, and that a straight line is 180°, and that any parallelogram contains 360°. So have some fun and figure it out.

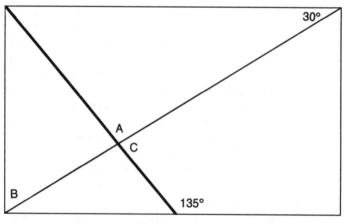

Fig. 5. A puzzler. From the information given in this figure, find the number of degrees in angles A, B and C.

This is what a career in plumbing and pipe fitting is all about. It is not just wielding a wrench under a kitchen sink, although there can be a lot of that. It is not all mathematics, either, although there definitely will be a lot of that.

As the *Wheels of Learning* course points out,

> Mathematics is the most durable of all tools. It can construct intricate piping systems, regardless of size, weight or materials. It can be carried anywhere effortlessly. It is guaranteed never to break, chip or crack for the life of the universe. It performs the same in any weather and under all conditions. Should it become slightly dulled, it can easily be restored to its former luster by rubbing pencil on paper. Amazingly, it costs nothing. Most importantly, it is the only tool that becomes sharper with use.

This continual exercise of both body and mind should be exciting for anyone planning now to shape a career for the future.

VGM CAREER BOOKS

OPPORTUNITIES IN

*Available in both
paperback and hardbound
editions*

Accounting Careers
Acting Careers
Advertising Careers
Agriculture Careers
Airline Careers
Animal and Pet Care
Appraising Valuation Science
Architecture
Automotive Service
Banking
Beauty Culture
Biological Sciences
Book Publishing Careers
Broadcasting Careers
Building Construction Trades
Business Communication Careers
Business Management
Cable Television
Carpentry Careers
Chemical Engineering
Chemistry Careers
Child Care Careers
Chiropractic Health Care
Civil Engineering Careers
Commercial Art and Graphic
 Design
Computer Aided Design
 and Computer Aided Mfg.
Computer Maintenance Careers
Computer Science Careers
Counseling & Development
Crafts Careers
Dance
Data Processing Careers
Dental Care
Drafting Careers
Electrical Trades
Electronic and Electrical
 Engineering
Energy Careers
Engineering Technology Careers
Environmental Careers
Fashion Careers
Federal Government Careers
Film Careers
Financial Careers
Fire Protection Services
Fitness Careers
Food Services
Foreign Language Careers
Forestry Careers
Gerontology Careers
Government Service
Graphic Communications

Health and
 Medical Careers
High Tech Careers
Home Economics Careers
Hospital Administration
Hotel & Motel Management
Industrial Design
Insurance Careers
Interior Design
International Business
Journalism Careers
Landscape Architecture
Laser Technology
Law Careers
Law Enforcement and
 Criminal Justice
Library and Information
 Science
Machine Trades
Magazine Publishing Careers
Management
Marine & Maritime Careers
Marketing Careers
Materials Science
Mechanical Engineering
Microelectronics
Modeling Careers
Music Careers
Nursing Careers
Nutrition Careers
Occupational Therapy
 Careers
Office Occupations
Opticianry
Optometry
Packaging Science
Paralegal Careers
Paramedical Careers
Part-time & Summer Jobs
Personnel Management
Pharmacy Careers
Photography
Physical Therapy Careers
Plumbing & Pipe Fitting
Podiatric Medicine
Printing Careers
Psychiatry
Psychology
Public Health Careers
Public Relations Careers
Real Estate
Recreation and Leisure
Refrigeration and
 Air Conditioning
Religious Service
Retailing
Robotics Careers
Sales Careers

Sales & Marketing
Secretarial Careers
Securities Industry
Social Work Careers
Speech-Language Pathology
 Careers
Sports & Athletics
Sports Medicine
State and Local Government
Teaching Careers
Technical Communications
Telecommunications
Television and Video Careers
Theatrical Design
 & Production
Transportation Careers
Travel Careers
Veterinary Medicine Careers
Vocational and Technical Careers
Word Processing
Writing Careers
Your Own Service Business

CAREERS IN

Accounting
Business
Communications
Computers
Health Care
Science

CAREER DIRECTORIES

Careers Encyclopedia
Occupational Outlook Handbook

CAREER PLANNING

How to Get and Get Ahead
 On Your First Job
How to Get People to Do
 Things Your Way
How to Have a Winning
 Job Interview
How to Land a Better Job
How to Write a Winning Résumé
Joyce Lain Kennedy's Career Book
Life Plan
Planning Your Career Change
Planning Your Career of
 Tomorrow
Planning Your College Education
Planning Your Military Career
Planning Your Own Home
 Business
Planning Your Young Child's
 Education

SURVIVAL GUIDES

High School Survival Guide
College Survival Guide

 **VGM Career Horizons**
a division of *NTC Publishing Group*
4255 West Touhy Avenue
Lincolnwood, Illinois 60646 1975

EDUCATION